INFOGRAPHIC

GUIDE TO

LIFE, THE UNIVERSE AND EVERYTHING

An Hachette UK Company
www.hachette.co.uk

First published in Great Britain in 2014 by Cassell Illustrated
a division of Octopus Publishing Group Ltd

Endeavour House
189 Shaftesbury Avenue
London
WC2H 8JY

www.octopusbooks.co.uk
www.octopusbooksusa.com

Distributed in the US by
Hachette Book Group USA
237 Park Avenue
New York NY 10017 USA

Distributed in Canada by
Canadian Manda Group
664 Annette Street
Toronto, Ontario, Canada M6S 2C8

ISBN 978-1-844037-88-9

A CIP catalogue record for this book is available from the British Library

Printed and bound in China

1 3 5 7 9 10 8 6 4 2

INFOGRAPHIC

GUIDE TO

LIFE, THE UNIVERSE AND EVERYTHING

Thomas Eaton

CONTENTS

Introduction 8

Black Holes 10

Modern Life Is Dangerous 12

Grand Torino 14

Where Did Everybody Go? 16

Wave Hello 18

World Wide Web 20

Extreme Weather Watch 21

A Mighty Wind 22

The Sleep Of Success 24

Pump It Up! 26

What Are You, A Man, Or A Mouse? 28

Chemical Composition 30

The Evolution Of Language 32

The New Power Generation 34

Worth More Dead Than Alive 36

The Sun Has Got His Heat On 38

The 47-Hour Day 39

New Planet, New You 40

Circles of Life 42

Tiny Noisemakers 44

Up, Up And Away 46

High Anxiety 48

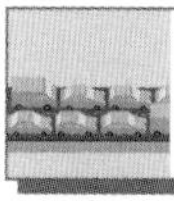
Crosstown Traffic Crazy 50

These Aren't The Planets You Are Looking For 52

Megacities Of The Future 54

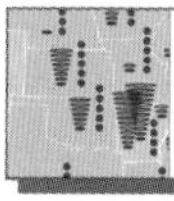
Not Only In Kansas, Toto 56

It's A Dog's Life Line 58

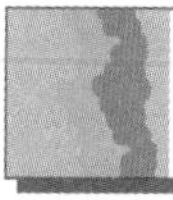
Throwing Air Bricks In A Greenhouse 60

Is It A Plane? Is It A Car? No, It's A Cow 61

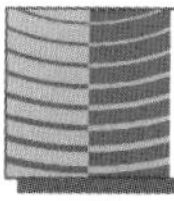
Shrimps Louder Than A Jet 62

Are We There Yet? 64

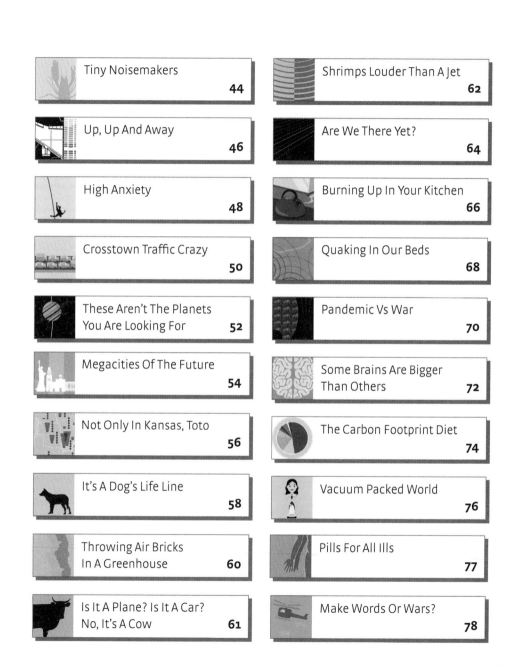

Burning Up In Your Kitchen 66

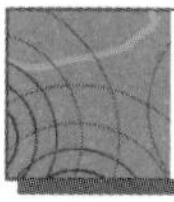
Quaking In Our Beds 68

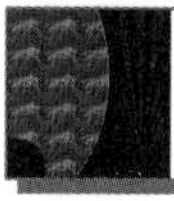
Pandemic Vs War 70

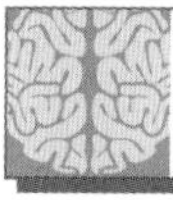
Some Brains Are Bigger Than Others 72

The Carbon Footprint Diet 74

Vacuum Packed World 76

Pills For All Ills 77

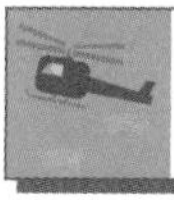
Make Words Or Wars? 78

Extreme Taste Test 80

How Low Can You Go? 82

Six Degrees Of Separation: Stephen Hawking 84

What's Out There 86

Lost In The Bermuda Triangle 88

Lost Tongues 90

Know Your Oligarch 92

All Creatures Great And Small 94

Boldly Going 96

Living Under A Volcano 98

Light Up The World 100

Elementary Nomenclature 102

What Planet Are You On? 104

Stamping Out Diseases 106

May Your God Go With You 108

Flatpack Forests 110

Abacus To Twitter 112

Climb Every Mountain 114

The Weight Of Happiness 116

Great Balls Of Fire 118

Who Really Runs The World 120

It'll Never Catch On 122

Alt, Shift... Delete? 124

Going Off The Grid 126

To Infinity... And Beyond? 128

Life, Death And Wealth 130

Inside The Large Hadron Collider 132

Snarling Up Traffic 133

Happy Accidents 134

Going Up And Down 136

Qualified To Run A Country? 138

World Of Dinosaurs 140

Snails Or Chocolate? 142

Everyone Who Ever Lived 144

Voyager In My Pocket 146

At The Edge Of The World 148

The Air That I Breathe 150

Water, Water, Everywhere 152

Gangster Universe 154

Work, Rest Or Play 156

Six Degrees Of Separation: Peter Higgs 158

Credits 160

Introduction

by Thomas Eaton

In 1834, a marble tablet was erected in St Andrew's chapel in Westminster Abbey, London, to commemorate a remarkable man called Thomas Young who had died five years before. The inscription described him as 'eminent in almost every department of human learning. Who, equally distinguished in the most abstruse investigations of letters and of science, first established the undulatory theory of light and first penetrated the obscurity which had veiled for ages the hieroglyphicks of Egypt'. A biography of this polymath was entitled The Last Man Who Knew Everything.

Who could even come close to this now? Even today's 'brainiacs' like Stephen Hawking, Noam Chomsky or Steven Pinker wouldn't claim to be experts in more than a handful of fields. We live in an age of the unparalleled growth of information; in all areas of knowledge the explosion of data is overwhelming. The World Wide Web has expanded from the few pages shared by Tim Berners-Lee and his colleagues at CERN in 1989 to a tool used by billions each week. There is so much information out there that scientists are running out of ways to describe it: terabytes were surpassed by petabytes, which grew into exabytes, then zettabytes and now yottabytes (that's 1 followed by 24 zeroes, if you're counting).

So how do we begin to process all this data and pick out what's most important? This book tries to do just that, with an infographical guide to life, the universe and everything. This way of representing information has a long and surprising tradition. Florence Nightingale is popularly remembered as the 'lady with the lamp', nursing wounded British troops in the Crimean War, but she was also a pioneering statistician. Her report *Mortality of the British Army* made extensive use of bar charts and innovative 'rose' diagrams to illustrate her findings 'by conveying ideas on the subject through the eye, which cannot be so readily grasped when contained in figures'. It earned her election in 1858 as the first female member of the Statistical Society.

The data is certainly out there. The question is how do we bring it all to life? This book covers an eclectic mix of topics, from sub-atomic particles travelling near the speed of light inside the Large Hadron Collider to the biggest stars in the universe. We've cast the net wide. We look in depth at space travel and the planets of our solar system – how long would it take you to get to Mars, and how much would you weigh when you got there? What are the chances of an asteroid strike wiping out life on Earth? It's happened before, you know... And then we venture out into deep space and even into the realms of science fiction.

Closer to home, there are pages devoted to the extremes of weather, volcanic activity, earthquakes and our planet's fragile climate. We show where the world's water is located, what's producing greenhouse gases and the carbon footprints left by different diets. You can discover how genetic make-up and relative brain size vary from species to species and how popular dog breeds developed. We break the human body down to its chemical bare essentials and see how successful we've been in eradicating diseases.

The book looks at modern life in all its dazzling technological complexity. What are the world's favourite websites? There's much more to the Internet than Google, Facebook and YouTube. How is the structure of the global population changing and how fast are the new megacities expanding? Where are the greatest concentrations of natural light? If this peek into the future is a bit daunting, there's even a guide to going off grid and getting away from it all.

Under the description of 'Everything' you'll find slightly more esoteric data illustrated, telling you which criminal gangs in which countries are the largest, oldest, and what they trade in, for instance. There's also a handy guide to Russian oligarchs, a comparison of the academic qualifications of the leaders of different nations, what's been lost in the Bermuda Triangle, and the nations who are heaviest and happiest.

In short, this book is a kaleidoscope of comparisons; some are serious, others decidedly not. It offers snapshots of the dizzying array of data out there. So dive in and remember that those frock-coated Victorian know-it-alls would give their right arm for a fraction of the knowledge available to us at the click of a mouse.

BLACK **HOLES**

Black holes have been observed at the centers of many galaxies beyond our own. It is believed that every galaxy has a black hole at its centre. A black hole's gravitational pull, felt in small degrees even at great distances, plays an integral role in shaping a galaxy.

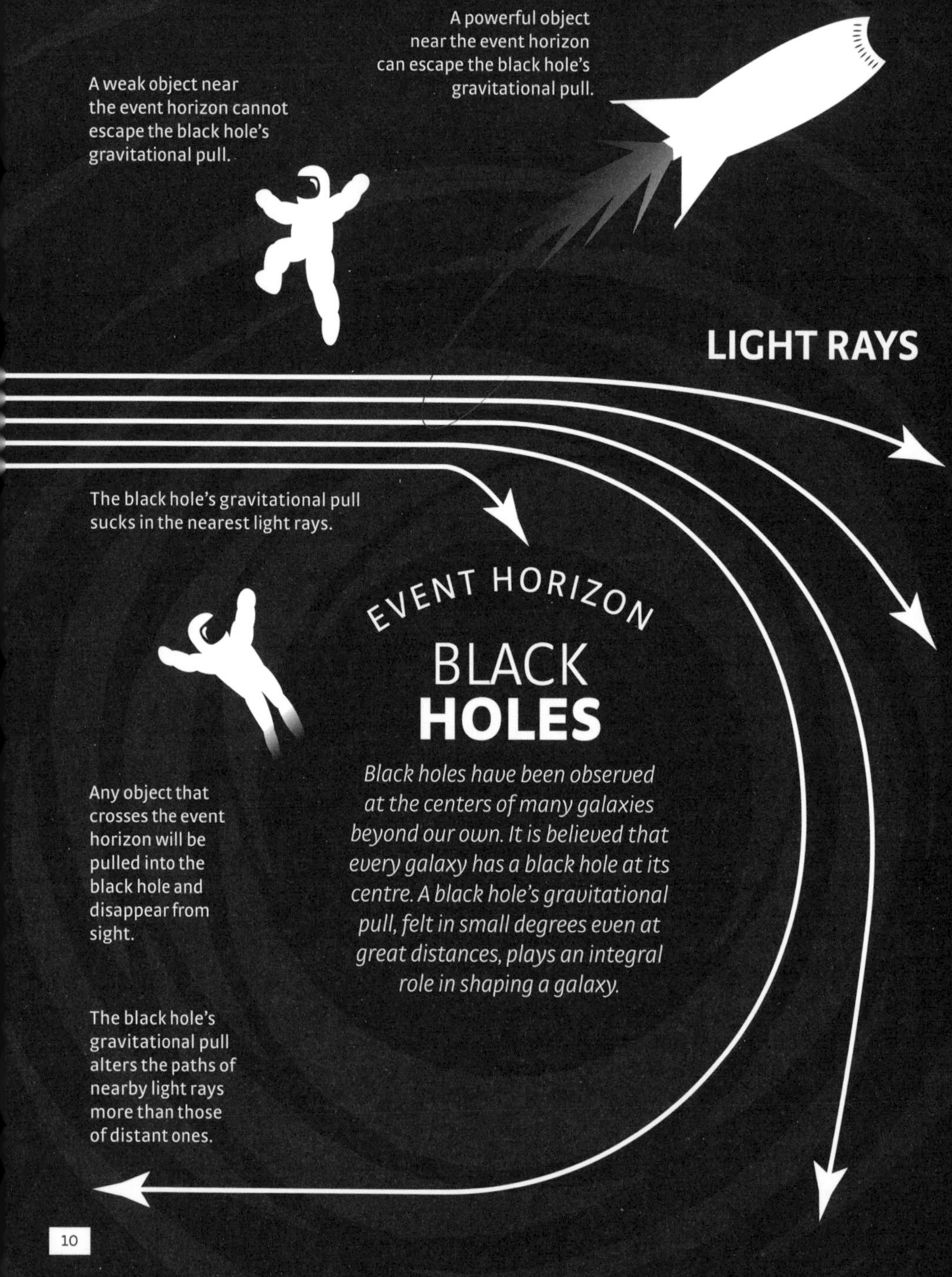

A powerful object near the event horizon can escape the black hole's gravitational pull.
A weak object near the event horizon cannot escape the black hole's gravitational pull.
LIGHT RAYS
The black hole's gravitational pull sucks in the nearest light rays.
EVENT HORIZON
Any object that crosses the event horizon will be pulled into the black hole and disappear from sight.
The black hole's gravitational pull alters the paths of nearby light rays more than those of distant ones.

MILKY WAY GALAXY

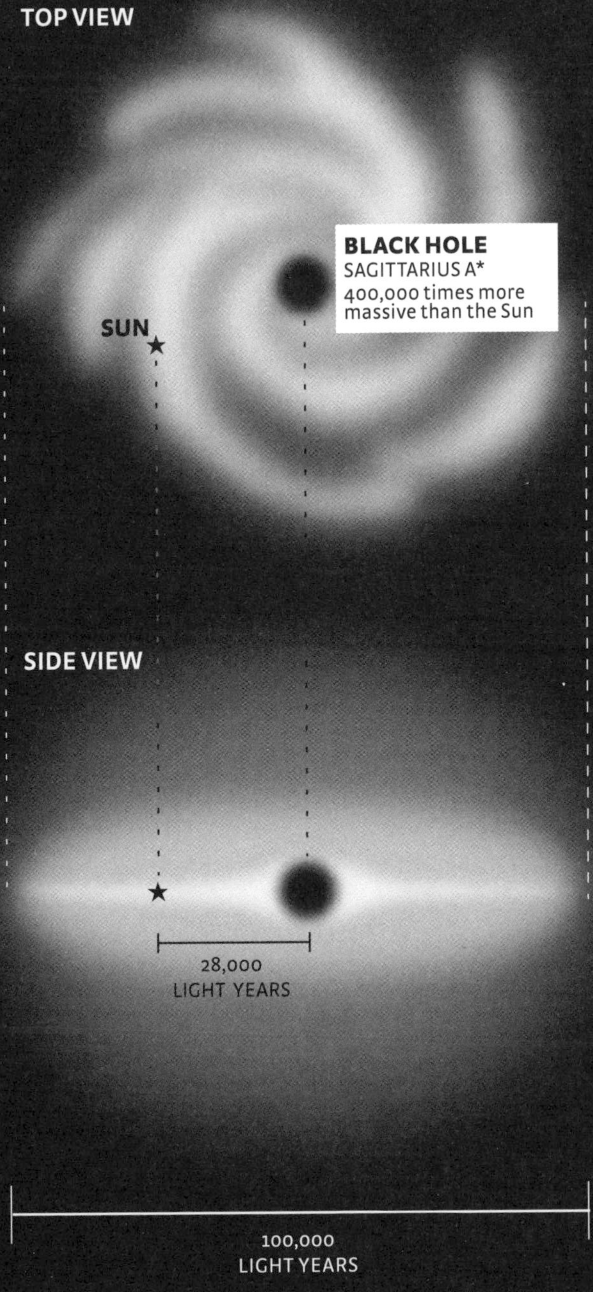

How has our understanding of **BLACK HOLES** developed?

1783 The idea of a body so massive that even light could not escape it is first put forward by geologist John Mitchell. It is largely ignored, since it is not yet understood how gravity can influence light.

1915 Albert Einstein develops his theory of general relativity, having shown that gravity influences light's motion.

1931 Subrahmanyan Chandrasekhar calculates that a black hole must have a radius of zero.

1958 David Finkelstein identifies the black hole's mathematical boundary, the event horizon.

1967 Physicist John Wheeler coins the term "black hole" during a lecture.

1974 Astronomers discover a supermassive black hole at the center of the Milky Way Galaxy.

2004 Stephen Hawking presents his new theory that black holes may allow information within them to escape. He theorizes that they keep emitting radiation for a long time and eventually open up to reveal the information within.

MODERN LIFE IS **DANGEROUS**

Despite the best practices of medicine, health & safety departments and education, men and women are killed by everyday items that we all live with. You're safer being female than male, as this survey of causes of deaths in a European country over a 4-year period shows.

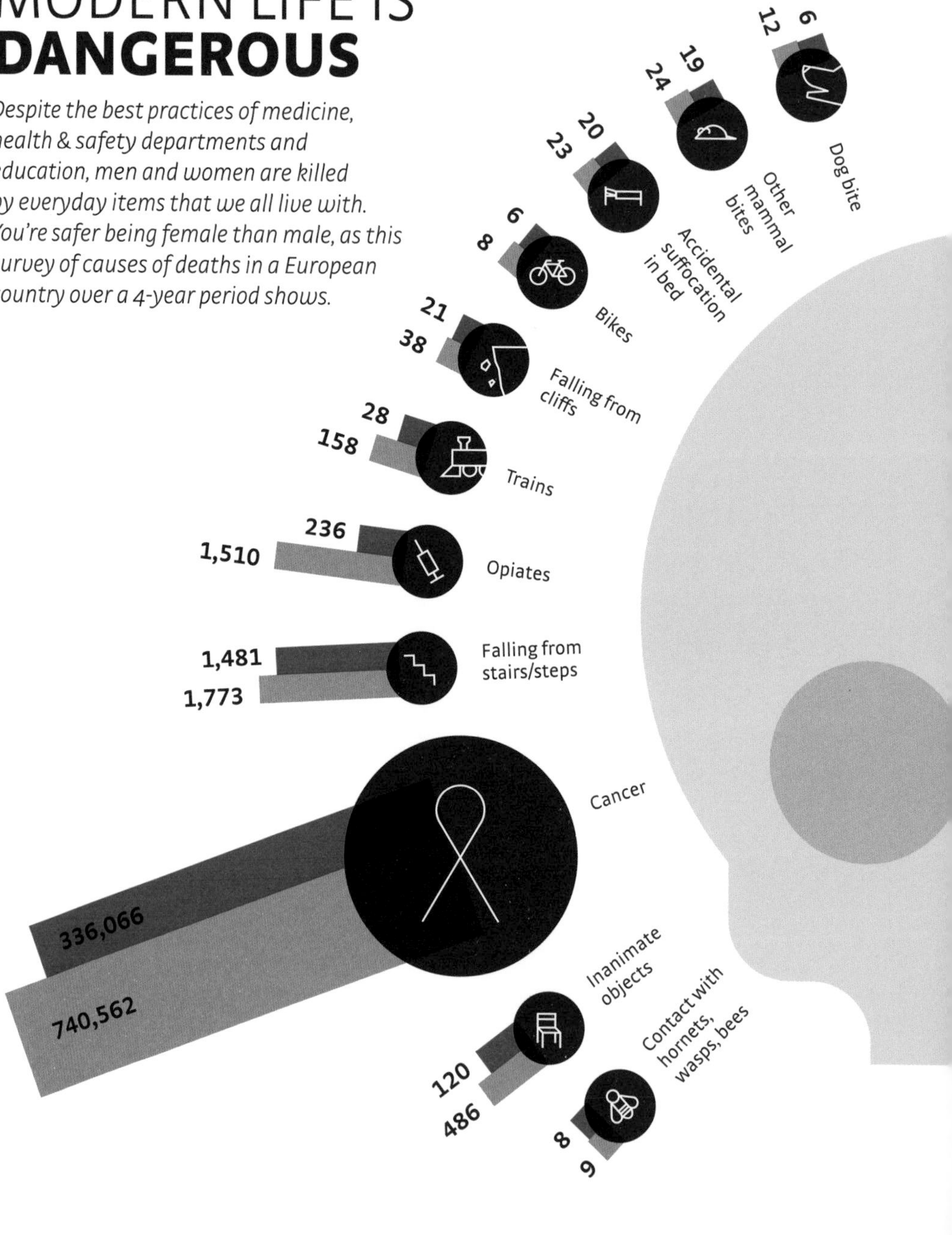

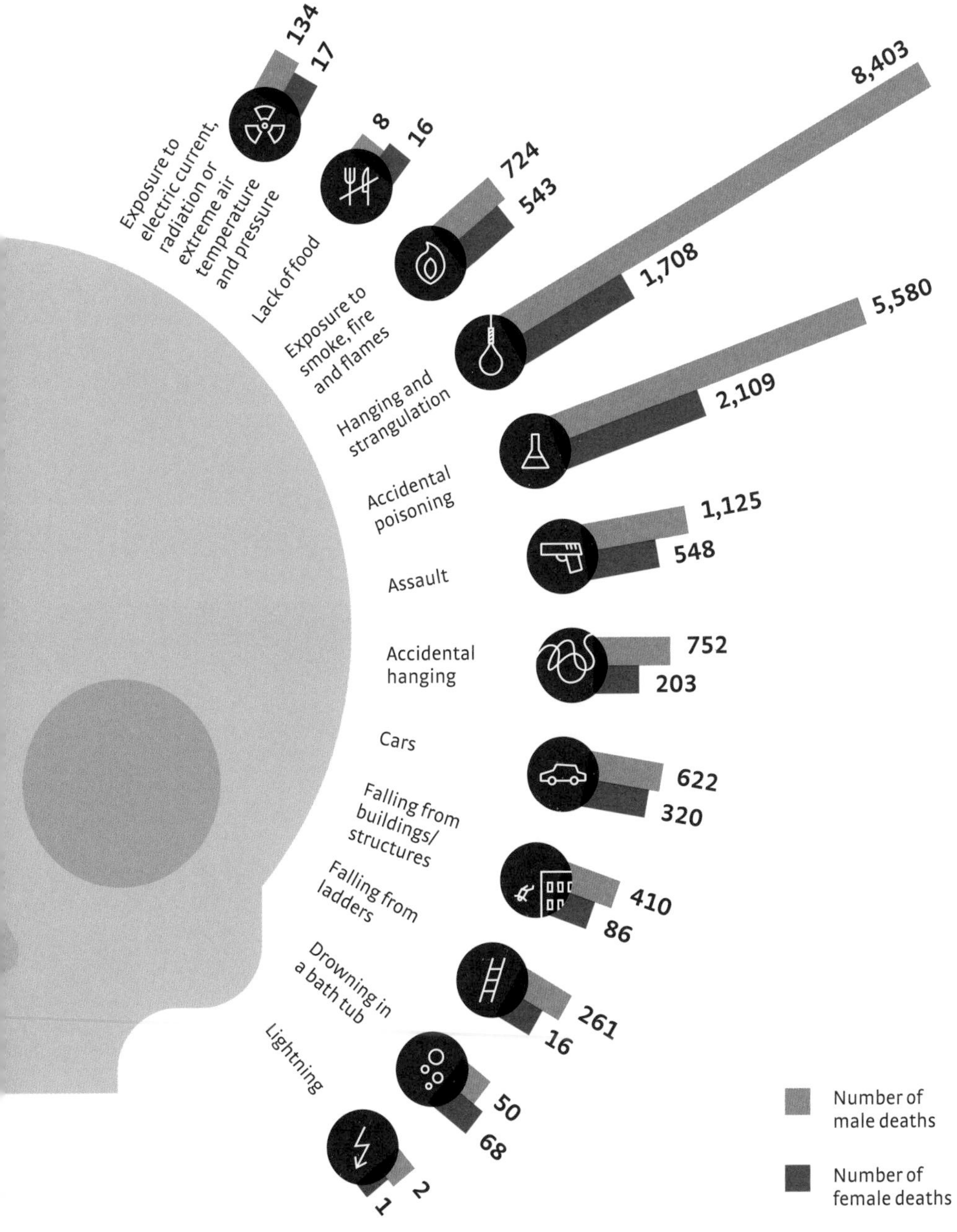

Exposure to electric current, radiation or extreme air temperature and pressure
134
17
Lack of food
8
16
Exposure to smoke, fire and flames
724
543
Hanging and strangulation
8,403
1,708
Accidental poisoning
5,580
2,109
Assault
1,125
548
Accidental hanging
752
203
Cars
622
320
Falling from buildings/ structures
410
86
Falling from ladders
261
16
Drowning in a bath tub
50
68
Lightning
2
1
Number of male deaths
Number of female deaths

GRAND **TORINO**

Our planet is constantly bombarded from space by particles the size of grains of sand. Objects as big as a small car hit the Earth a few times each year. NASA calculates that there are over a thousand near-Earth objects more than a kilometer in diameter and more than a million around the 40-meter mark. Astronomers use the Torino Impact Hazard Scale to assess the potential danger posed by the impact of near-Earth objects.

No hazard (white)

0 The likelihood of the object colliding with Earth is zero, or virtually zero.

Normal (green)

1 The object will pass near Earth, but is predicted to cause no unusual level of danger. NASA's Sentry Risk Table of potential future collisions lists just one object, known as 2007 VK184, at risk level 1 on the Torino scale. This near-Earth asteroid is estimated to be 130 meters wide and is thought to have a 0.055% chance of hitting Earth on 3 June 2048.

Meriting attention by astronomers (yellow)

2 The discovered object is predicted to make a pass near Earth, which may become routine with more research. New observations are very likely to reassign its risk as zero.

3 A close encounter, meriting further investigation. Current information suggests a 1% or greater chance of collision capable of localized destruction. Telescopic observations will very likely reduce the risk level to zero.

4 A 1% or greater chance of a collision capable of regional devastation. The risk level will most likely drop to zero after further observation. In 2004, the asteroid 99942 Apophis was given a Torino scale rating of 4 because of an estimated 2.7% chance it would hit Earth in 2029. At around 330 meters in diameter, it was thought Apophis would enter the atmosphere with 750 megatons of kinetic energy (compared with the 50+ megatons from the biggest hydrogen bomb detonations). However, subsequent observations have led to its Torino rating being reduced to zero. So far no other object has been given a higher rating than 4.

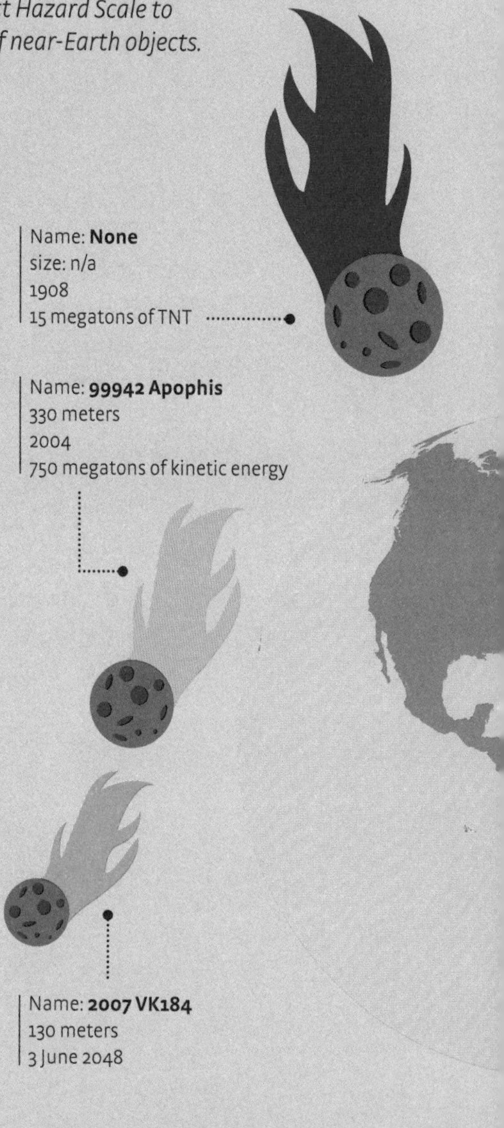

The Chicxulub crater – Mexico's Yucatán Peninsula – thought to have led to the mass extinction at the end of the Cretaceous Period – 65 million years ago

Threatening (orange)

5 A close encounter posing a serious, but still uncertain, threat of regional devastation. Astronomers must urgently determine whether or not it will collide with Earth. If the encounter is less than a decade away, governments might start planning emergency measures.

6 A large object posing a serious, but still uncertain, threat of a global catastrophe. Government emergency planning warranted if the potential collision is less than 30 years away.

7 A very close encounter by a large object. If it occurs within 100 years it poses a great, but still uncertain, threat of global catastrophe.

Certain collisions (red)

8 A collision is certain, capable of causing localized destruction. The asteroid or comet that collided with Tunguska in Siberia in 1908 was estimated to have realized energy equivalent to 15 megatons of TNT (a thousand times more powerful than the atomic bomb dropped on Hiroshima). The energy of the impact and the devastation over an area 60km across would have placed this, retrospectively, at 8 on the Torino scale.

9 A collision is certain, causing potentially unprecedented regional devastation.

10 Collision is certain, capable of causing global catastrophe. It may threaten the future of civilization on Earth. For example, the Chicxulub crater, off Mexico's Yucatán Peninsula, is the site of the impact thought to have led to the mass extinction at the end of the Cretaceous Period, around 65 million years.

Oxford Dic of Astronomy, Hutchinson Unabridged Enc, neo.jpl.nasa.gov/torino_scale.html

WHERE DID **EVERYBODY GO?**

As Cassandra said – and correctly so – we're all doomed. At least six times in the past a major catastrophe has befallen the earth with the result that a significant proportion of living creatures have perished. Can we learn anything from them?

Around **445** Million years ago

Event: Late Ordovician

Organisms affected: Over 90% of trilobite species became extinct

90%

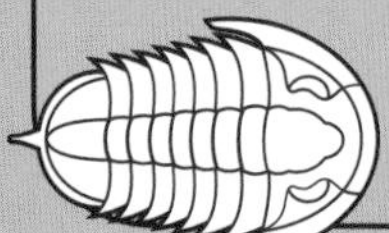

% of species extinct: around **75–85%**

Possible cause: Glaciation

Around **250** Million years ago

Event: Late Permian, or the Great Dying

% of species extinct: Around **95%** of all species

Possible cause: High global temperatures, volcanic eruptions and lack of oxygen in the sea

Around **65** Million years ago

Event: Cretaceous–Tertiary, or K–T extinction

Organisms affected: Non-avian dinosaurs, some mammals, marine lizards

% of species extinct: **75–85%**

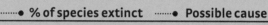

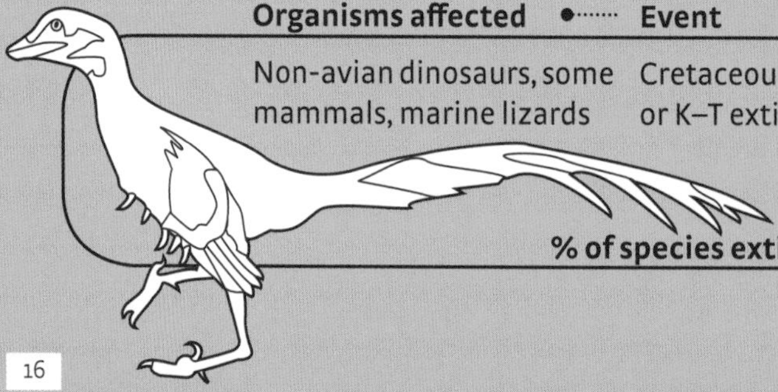

Around

488

Million years ago

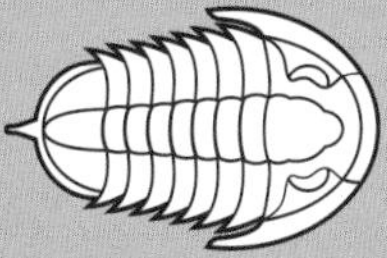

Event

Late Cambrian

Organisms affected

Sea creatures such as trilobites and brachiopods were greatly reduced in number

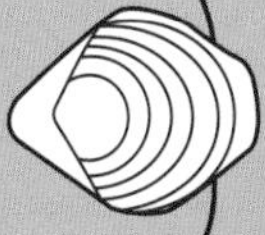

Possible cause

Changes in sea level

% of species extinct

No-one knows

Around

360

Million years ago

Event

Late Devonian

Organisms affected

Armoured fish died out entirely; many marine species were depleted

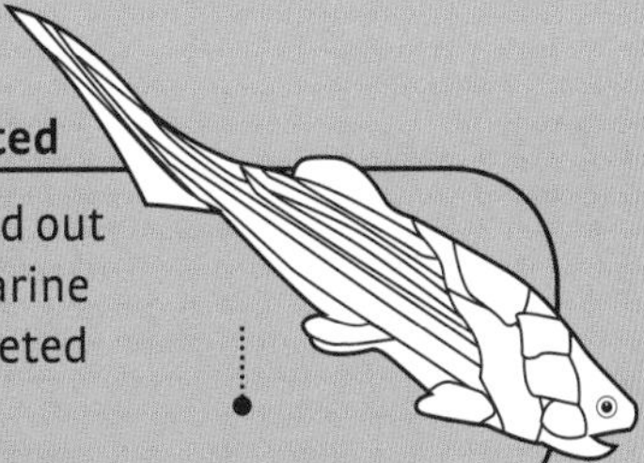

Possible cause

Reduced oxygen levels in water, rising sea levels and global cooling

% of species extinct

70–85% of all species

Around

200

Million years ago

Event

Late Triassic

Organisms affected

Almost all trilobites and many species of early insects

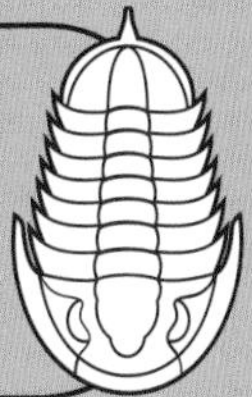

Possible cause

Climate change caused by continental drift

% of species extinct

80%

Possible cause

Meteorite or comet collision with Earth, leading to climate change

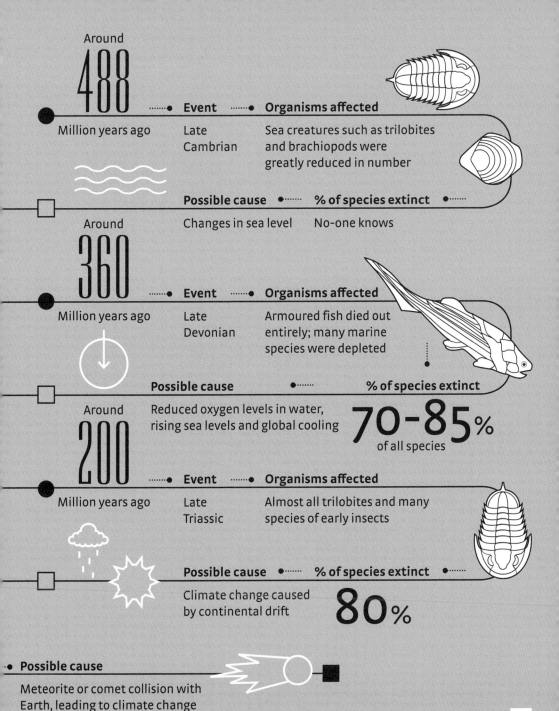

Oxford Dictionary of Biology, nhm.ac.uk

WAVE HELLO

Scientists have managed to lasso enough of the electromagnetic spectrum to make serious waves for mankind's benefit. Here are our favorite waves:

RADIO WAVES Radio waves have the longest wavelength on the electromagnetic spectrum. Used for broadcasting radio and TV, they are also found in the background radiation of the universe.

MICROWAVES Used in radar, communications, microwave ovens and cellphones.

INFRARED RAYS Emitted by hot objects, they have applications in communications, medicine, astronomy and the military.

VISIBLE LIGHT The electromagnetic radiation that is visible to the human eye.

ULTRAVIOLET LIGHT Most of the ultraviolet light in sunlight is filtered by the ozone layer. Prolonged exposure to UV rays can cause sunburn and even skin cancer.

X-RAYS Used for medical diagnosis and in airport security scanners.

GAMMA RAYS Used in the detection and treatment of some cancers and the sterilization of medical equipment. Also used in the food industry to kill off microorganisms and extend the shelf life of food.

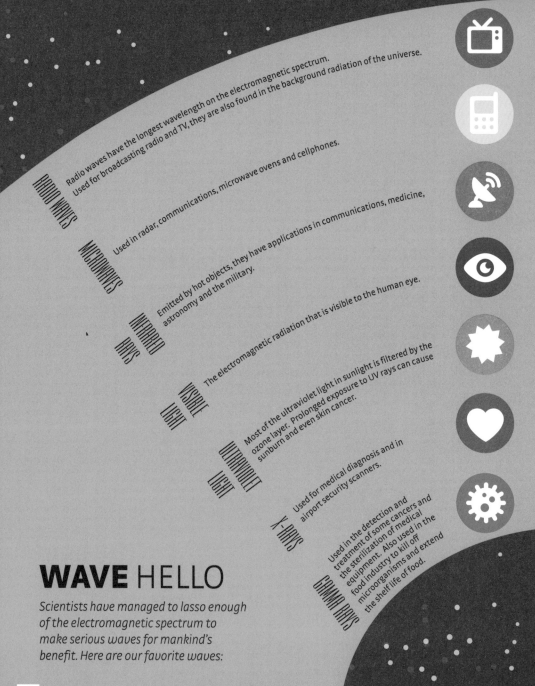

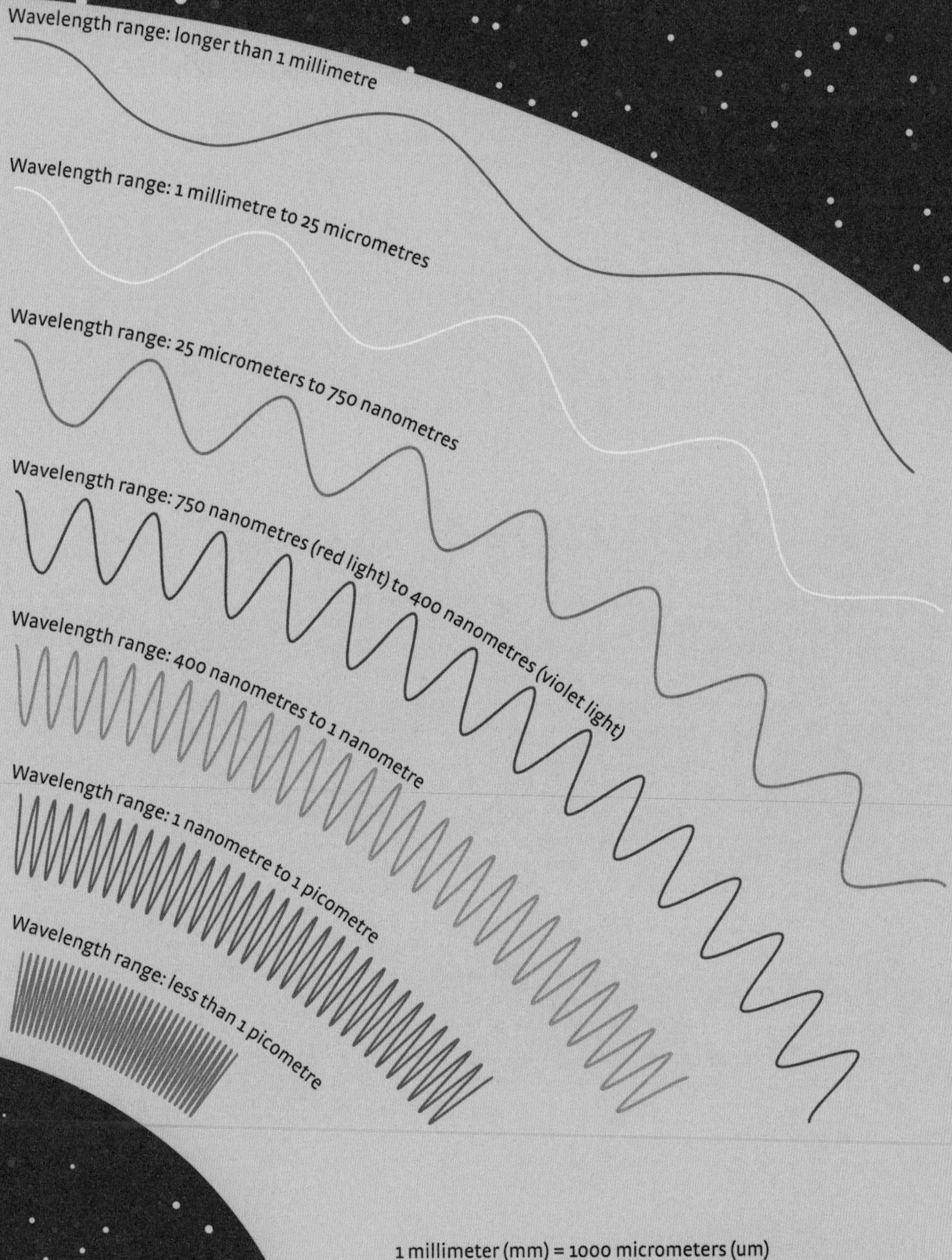

1 millimeter (mm) = 1000 micrometers (um)
1 micrometer (um) = 1000 nanometers (nm)
1 nanometer (nm) = 1000 picometers (pm)

WORLD WIDE **WEB**

What a tangled and widespread web we've woven, virtually speaking.

Global Internet Access (by % Households)

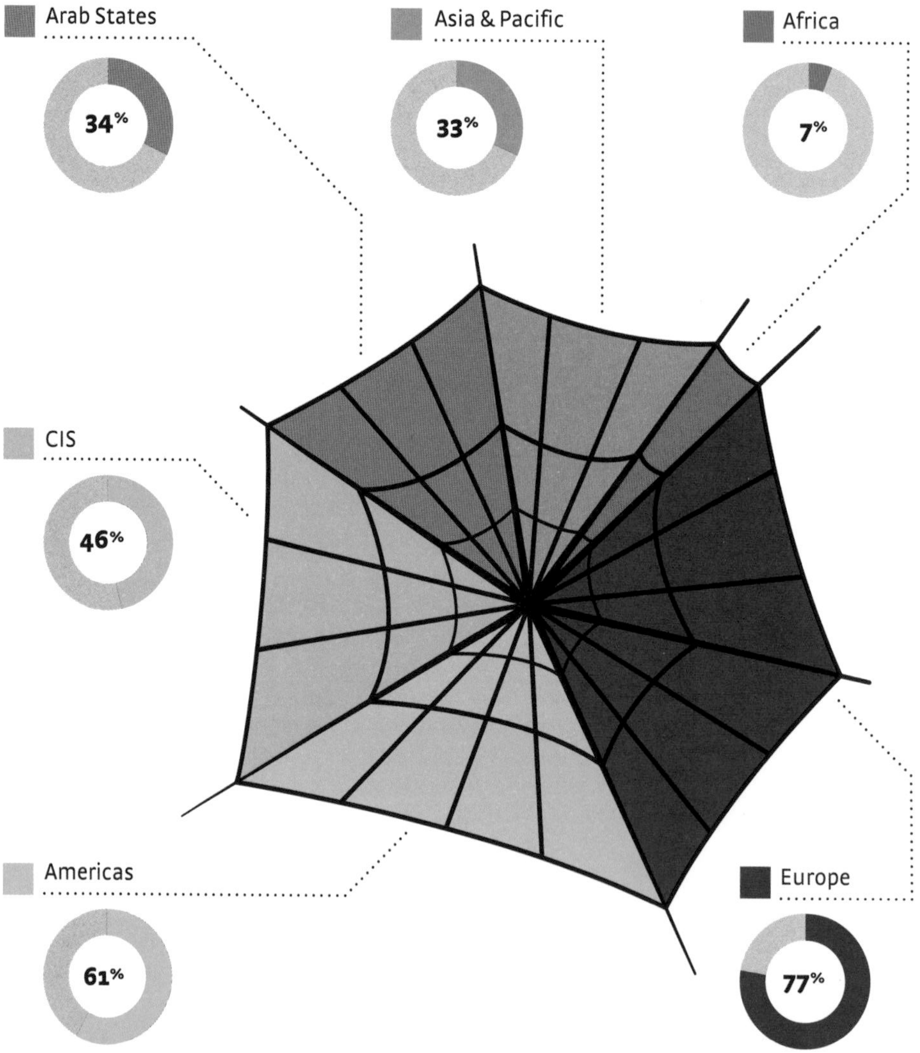

alexa.com, geography.oii.ox.ac.uk, itu.int

EXTREME
WEATHER WATCH

Whatever the weather's like where you live, check it against these places to see how lucky you are to live where you do. Or if you prefer it to be hotter, wetter, drier or colder, here's where to go.

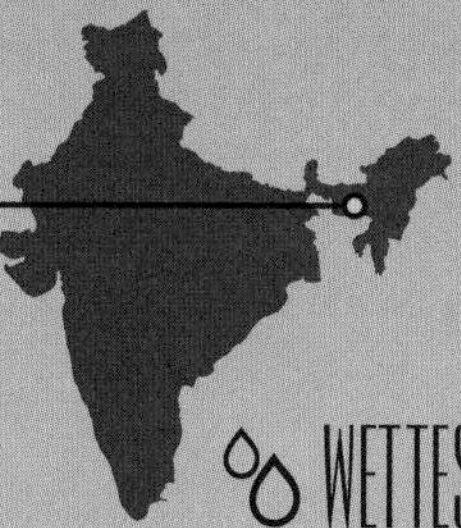

WETTEST

The village of Mawsynram, Meghalaya state in north-east India has an average annual rainfall of 11,873 millimeters.

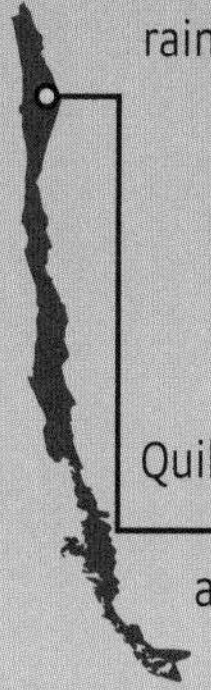

DRIEST

Quillagua, a village in Chile's Atacama Desert, has an annual average rainfall of around half a millimeter.

The driest place in the world is the Dry Valleys region of Antarctica, which gets virtually no rain or snowfall.

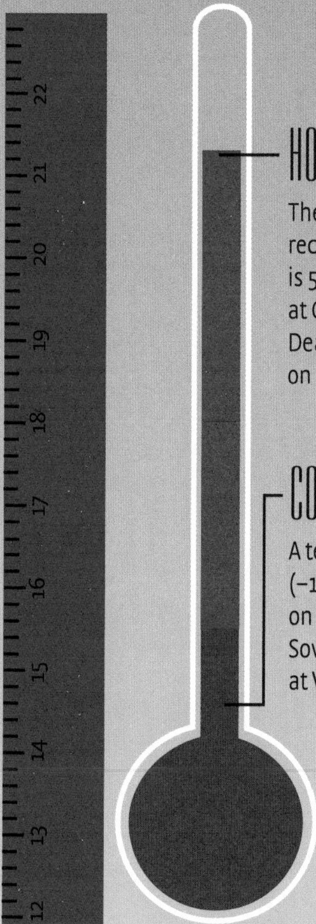

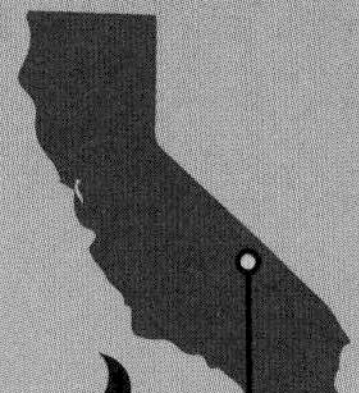

HOTTEST

The highest reliably recorded shade temperature is 56.7°C (134°F) recorded at Greenland Ranch in Death Valley, California on 10 July 1913.

COLDEST

A temperature of −89.2°C (−128.6°F) was recorded on 21 July 1983 at the Soviet research station at Vostok, Antarctica.

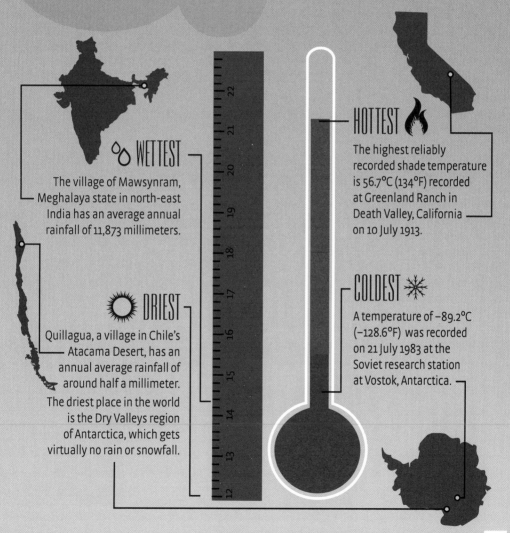

A MIGHTY **WIND**

When is a gale not a tornado, a tornado not a hurricane? When Messrs Beaufort, Fujita and Saffir-Simpson say so, that's when. And how fast is a tornado anyway? Here are the scales and some man-made speed limits to go with them.

Human Speed –

Wind Speed km/h -
>1 (1 mph)
1–5.5 (1–3 mph)
5.6–11 (4–7 mph)
12–19 (8–12 mph)
20–28 (13–17 mph)
29–38 (18–24 mph)
39–49 (25–30 mph)
50–61 (31–38 mph)
62–74 (39–46 mph)
75–88 (47–54 mph)
89–102 (55–63 mph)
103–117 (64–73 mph)
118+ (74 mph)
119–153 (75–95 mph)
154–177 (96–110 mph)
178–208 (111–129 mph)
209–251 (130–156 mph)
252+ (157 mph)
333–418 (207–259 mph)
419–512 (260–318 mph)

BEAUFORT Scale

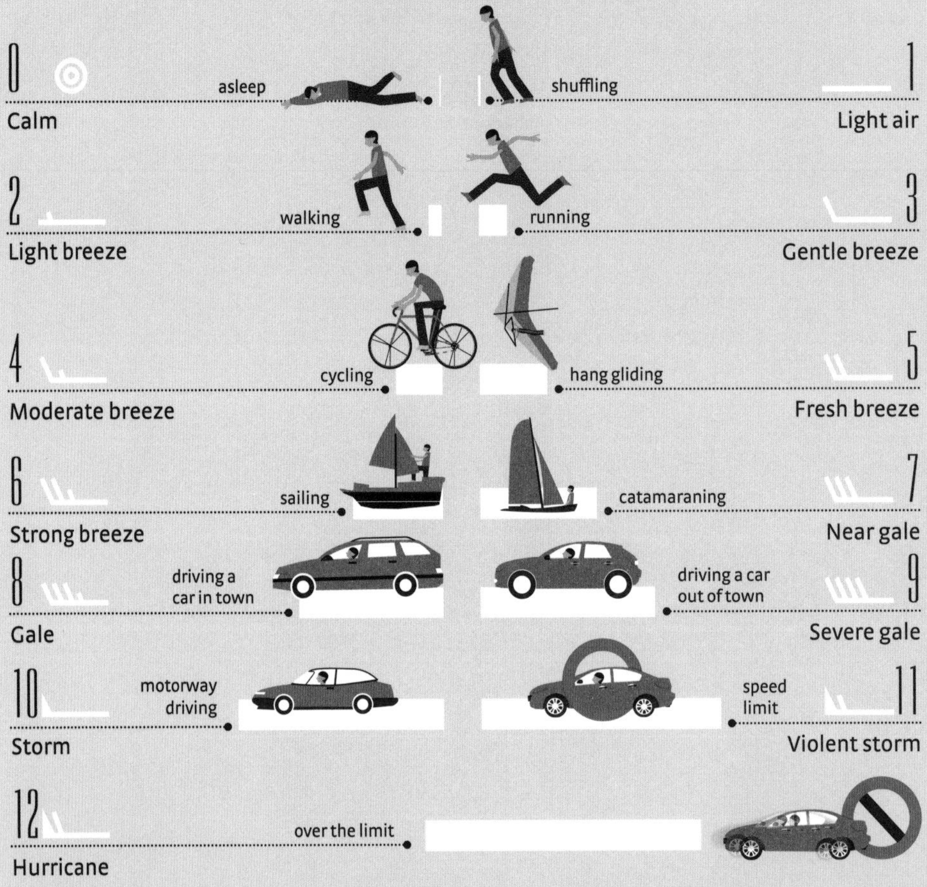

FUJITA Tornado Intensity Scale

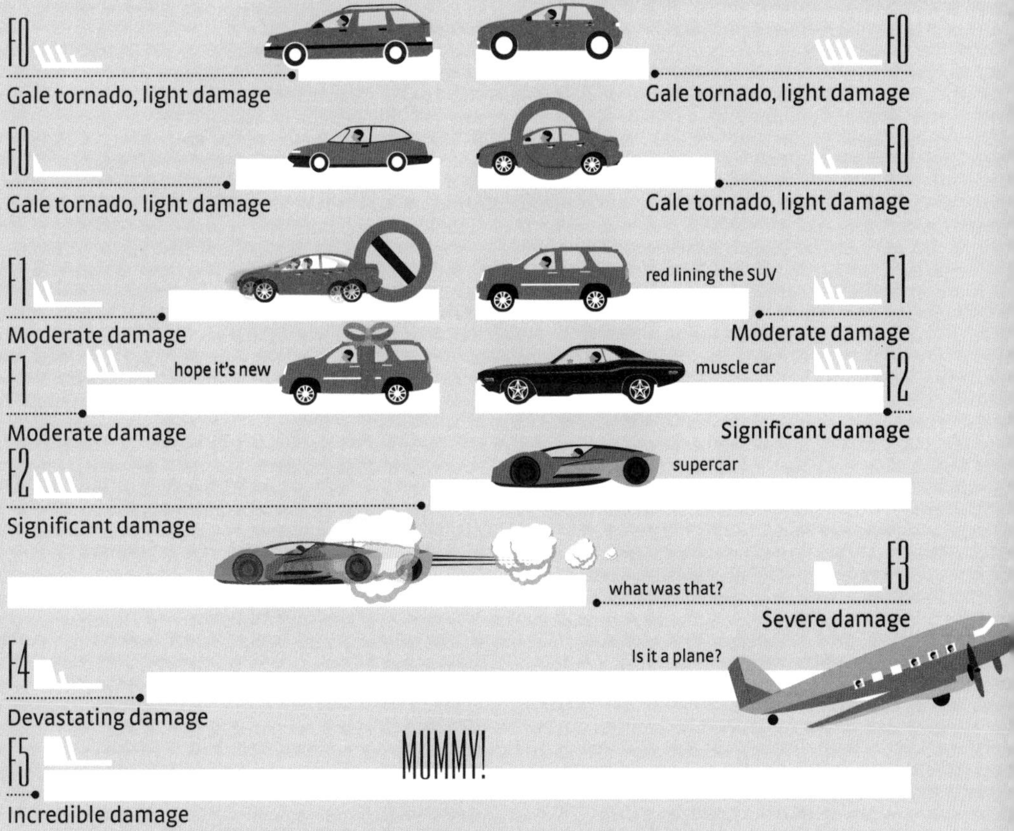

SAFFIR-SIMPSON Hurricane Wind Scale

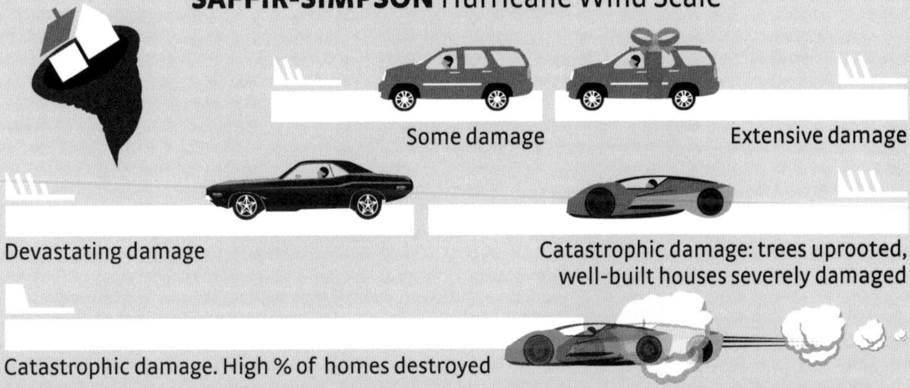

nhc.noaa.gov, metoffice.gov.uk, outlook.noaa.gov

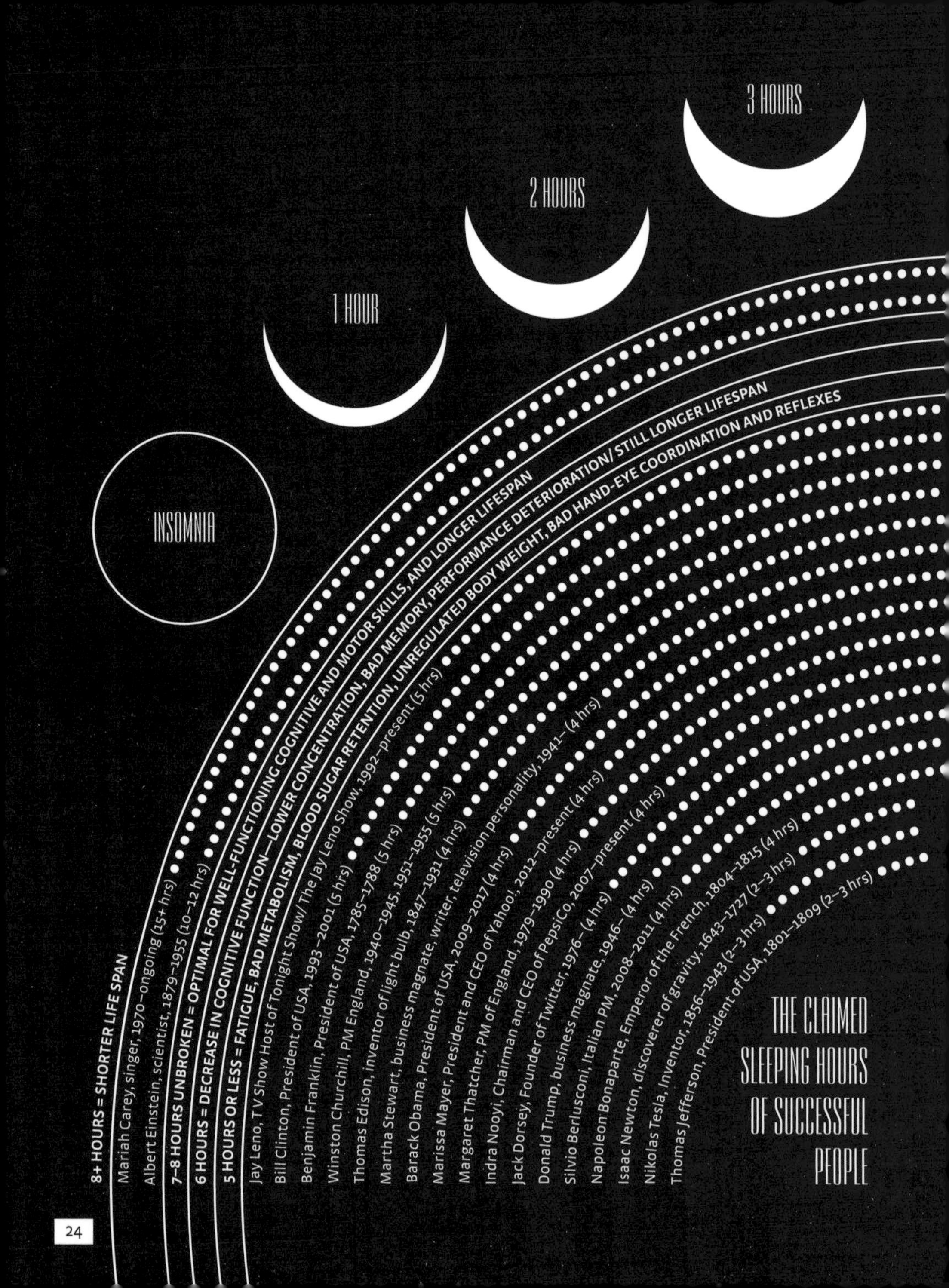
THE CLAIMED SLEEPING HOURS OF SUCCESSFUL PEOPLE
INSOMNIA
1 HOUR
2 HOURS
3 HOURS
8+ HOURS = SHORTER LIFE SPAN
Mariah Carey, singer, 1970–ongoing (15+ hrs)
Albert Einstein, scientist, 1879–1955 (10–12 hrs)
7–8 HOURS UNBROKEN = OPTIMAL FOR WELL-FUNCTIONING COGNITIVE AND MOTOR SKILLS, AND LONGER LIFESPAN
6 HOURS = DECREASE IN COGNITIVE FUNCTION—LOWER CONCENTRATION, BAD MEMORY, PERFORMANCE DETERIORATION/ STILL LONGER LIFESPAN
5 HOURS OR LESS = FATIGUE, BAD METABOLISM, BLOOD SUGAR RETENTION, UNREGULATED BODY WEIGHT, BAD HAND-EYE COORDINATION AND REFLEXES
Jay Leno, TV Show Host of Tonight Show/ The Jay Leno Show, 1992–present (5 hrs)
Bill Clinton, President of USA, 1993–2001 (5 hrs)
Benjamin Franklin, President of USA, 1785–1788 (5 hrs)
Winston Churchill, PM England, 1940–1945, 1951–1955 (5 hrs)
Thomas Edison, inventor of light bulb, 1847–1931 (4 hrs)
Martha Stewart, business magnate, writer, television personality, 1941– (4 hrs)
Barack Obama, President of USA, 2009–2017 (4 hrs)
Marissa Mayer, President and CEO of Yahoo!, 2012–present (4 hrs)
Margaret Thatcher, PM of England, 1979–1990 (4 hrs)
Indra Nooyi, Chairman and CEO of PepsiCo, 2007–present (4 hrs)
Jack Dorsey, Founder of Twitter, 1976– (4 hrs)
Donald Trump, business magnate, 1946– (4 hrs)
Silvio Berlusconi, Italian PM, 2008–2011 (4 hrs)
Napoleon Bonaparte, Emperor of the French, 1804–1815 (4 hrs)
Isaac Newton, discoverer of gravity, 1643–1727 (2–3 hrs)
Nikolas Tesla, Inventor, 1856–1943 (2–3 hrs)
Thomas Jefferson, President of USA, 1801–1809 (2–3 hrs)

THE SLEEP OF **SUCCESS**

History is littered with high achievers who boldly stated that sleep was for wimps. Scientific research in the 21st century, however, suggests that they were deluded, and for good reason, as is shown here.

wikipedia.org, sleepfoundation.org

PUMP IT **UP!**

Looking at how much oil is being extracted per day against each country's estimated reserves gives us an idea of how long we can be dependent on fossil fuels. The size of each country's reserves also tells us who will remain the richest and most powerful as long as we need their natural resources.

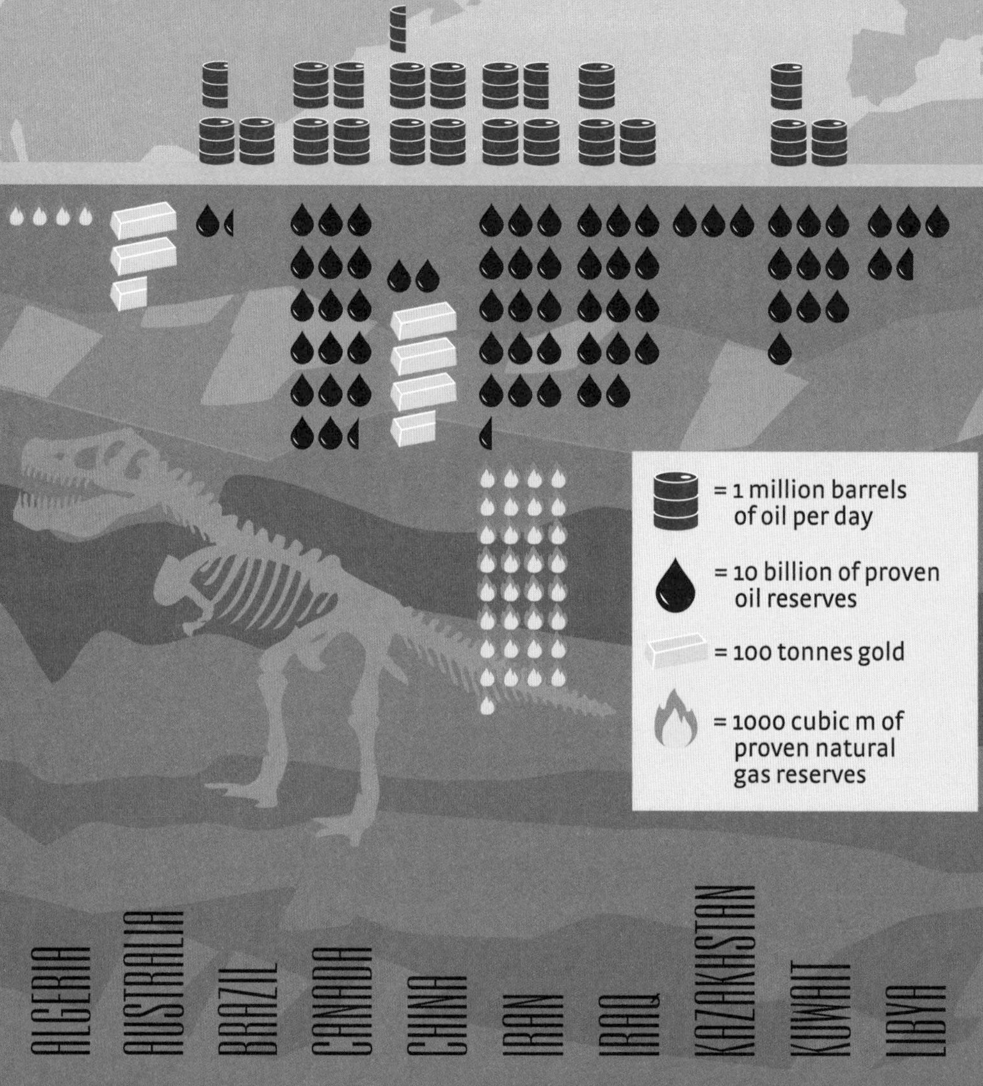

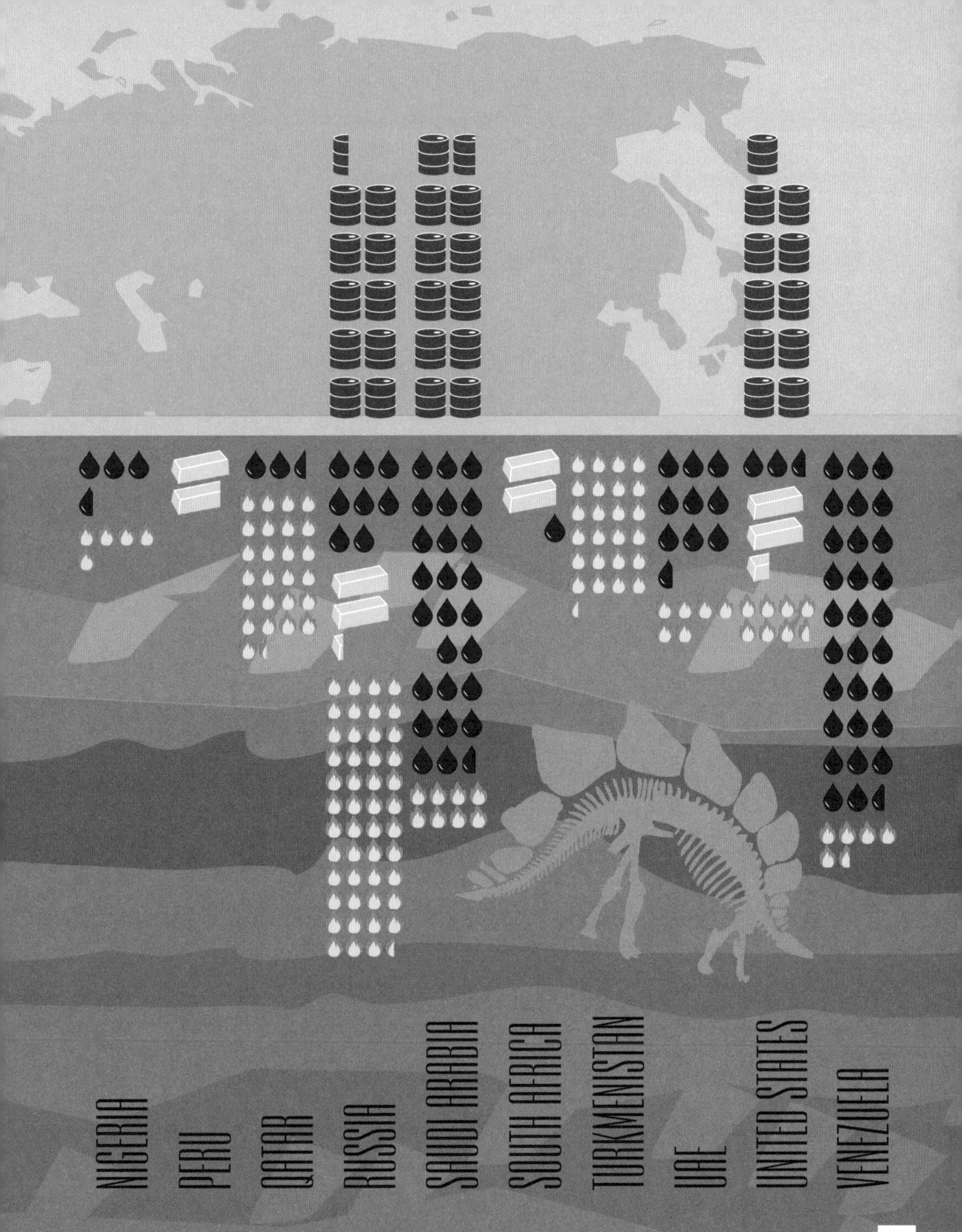

nhc.noaa.gov, metoffice.gov.uk, outlook.noaa.gov

WHAT ARE YOU, A MAN **OR A MOUSE?**

It's a close call actually. Human genetic makeup is only slightly more complicated than that of a fruit fly, and our genome size is remarkably smaller.

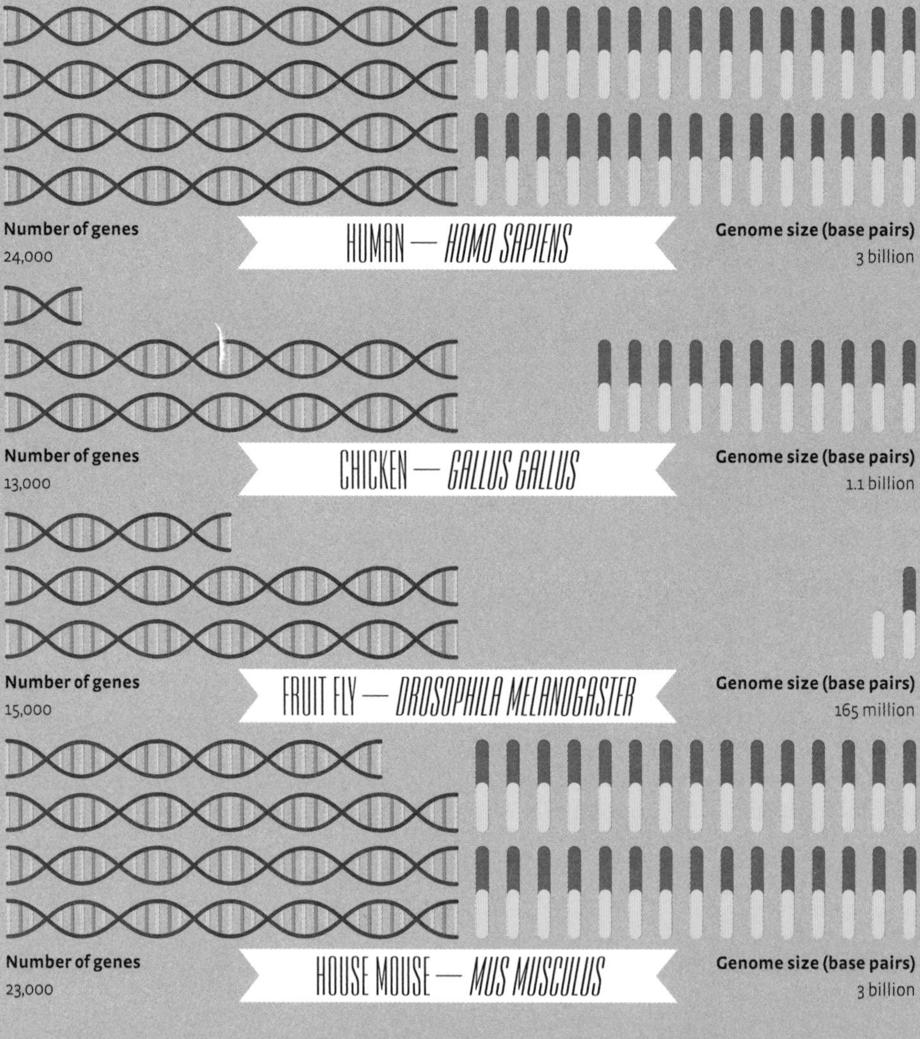

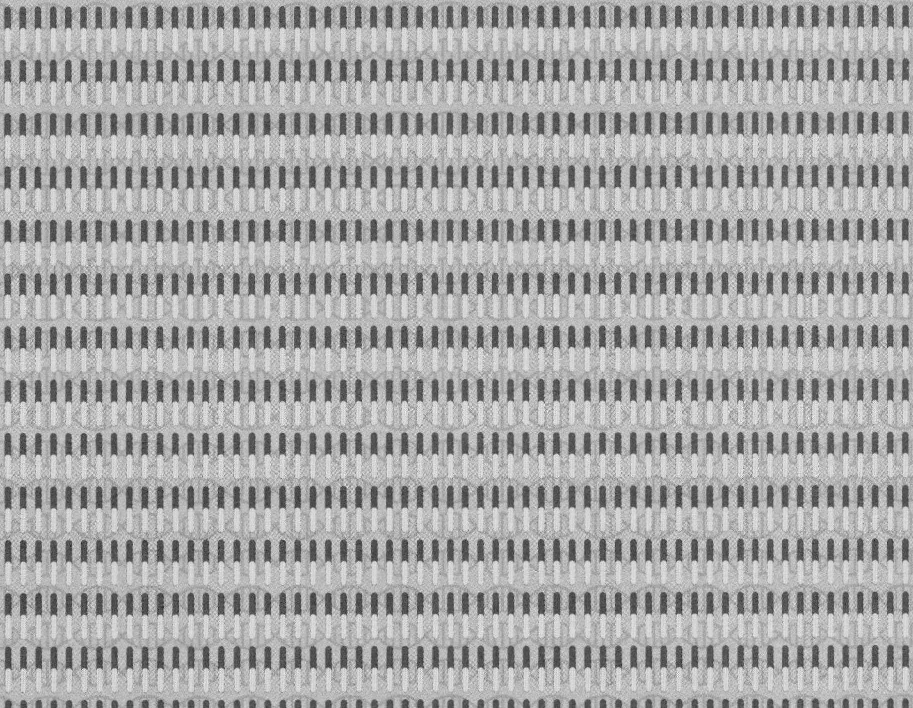

Number of genes not yet fully sequenced

MARBLED LUNGFISH — *PROTOPTERUS AETHIOPICUS*

Genome size (base pairs)
130 billion (largest known genome of any animal)

nature.com, Oxford Dictionary of Biology, ncbi.nlm.nih.gov/genome, genome.gov

CHEMICAL **COMPOSITION**

The average 70kg/154lb adult human contains approximately 7×10^{27} atoms and detectable traces of 60 chemical elements. The 28 elements shown here are thought to play an active positive role in life and health.

◆ Mass (kg)
○ Atomic percent

Atomic number	Symbol	Element	◆ Mass (kg)	○ Atomic percent
3	Li	Lithium	0.000007	0.0000015
17	Cl	Chlorine	0.095	0.024
6	C	Carbon	16	12
24	Cr	Chromium	0.000014	8.9E-08
12	Mg	Magnesium	0.019	0.007
7	N	Nitrogen	1.8	0.58
27	Co	Cobalt	0.000003	0.0000003
26	Fe	Iron	0.0042	0.00067
15	P	Phosphorus	.078	0.14
28	Ni	Nickel	0.000015	0.0000015
30	Zn	Zinc	0.0023	0.00031
16	S	Sulfur	0.14	0.038

			◆	○
8	O	Oxygen	43	24
9	F	Fluorine	0.0026	0.0012
53	I	Iodine	0.00002	0.00000075
42	Mo	Molybdenum	0.000005	4.5E-08
1	H	Hydrogen	7	63
14	Si	Silicon	0.001	0.0058
25	Mn	Manganese	0.000012	0.000015
20	Ca	Calcium	1	0.24
34	Se	Selenium	0.00014	4.5E-08
29	Cu	Copper	0.000072	0.0000104
19	K	Potassium	0.14	0.033
5	B	Boron	0.000018	0.000003
23	V	Vanadium	0.00000011	1.2E-08
11	Na	Sodium	0.1	0.037
33	As	Arsenic	0.000007	8.9E-08
35	Br	Bromine	0.00026	0.00003

Enc Britannica, Hutchinson's Unabridged Enc, foresight.org/Nanomedicine

THE EVOLUTION OF **LANGUAGE**

Man's ability to communicate began with grunts pre-10,000 years Before Common Era and by the 21st century has reached gangnam style and tweets. As far as experts can tell, before tribes started wandering all over the earth, they all spoke the same kind of proto-Indo-European. Then they began speaking in their own tongues.

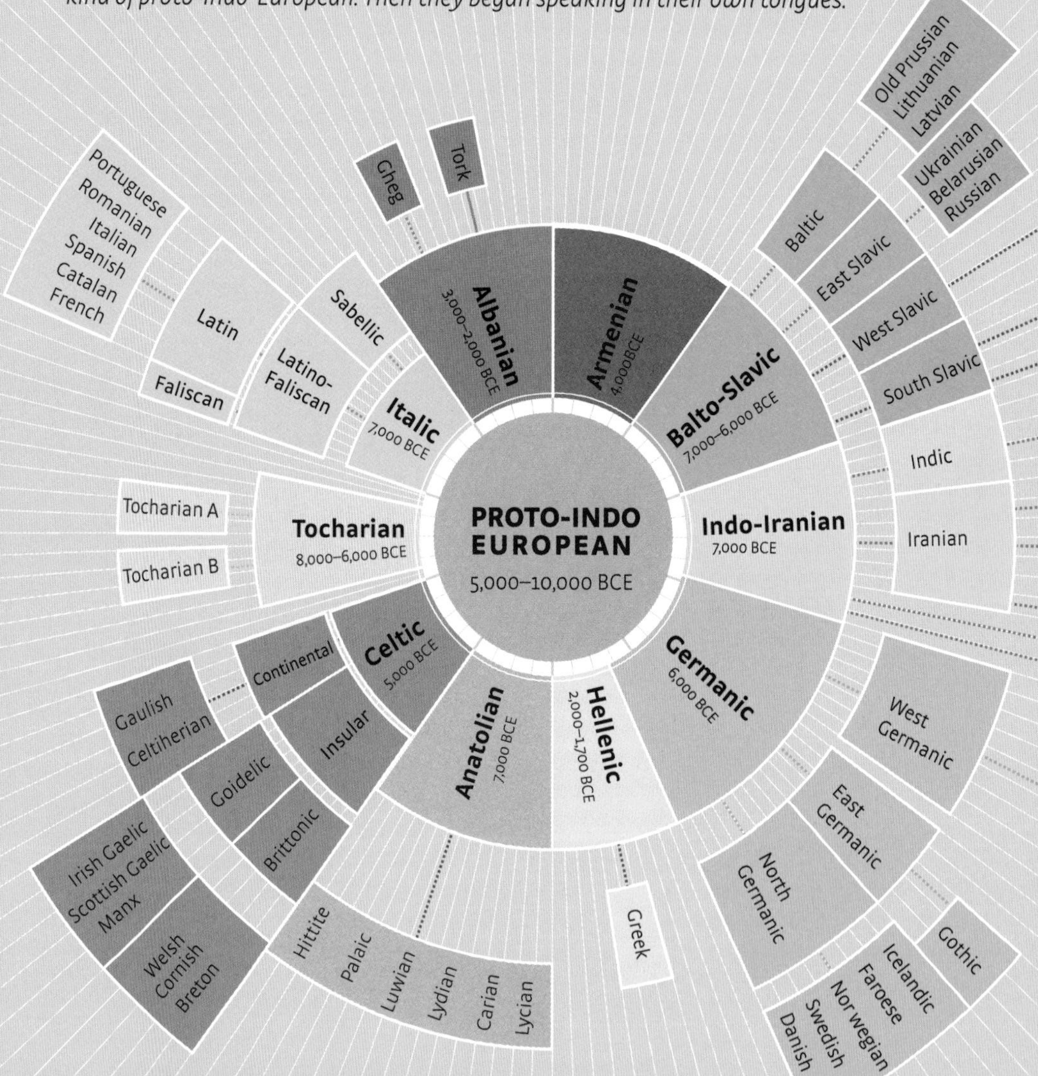

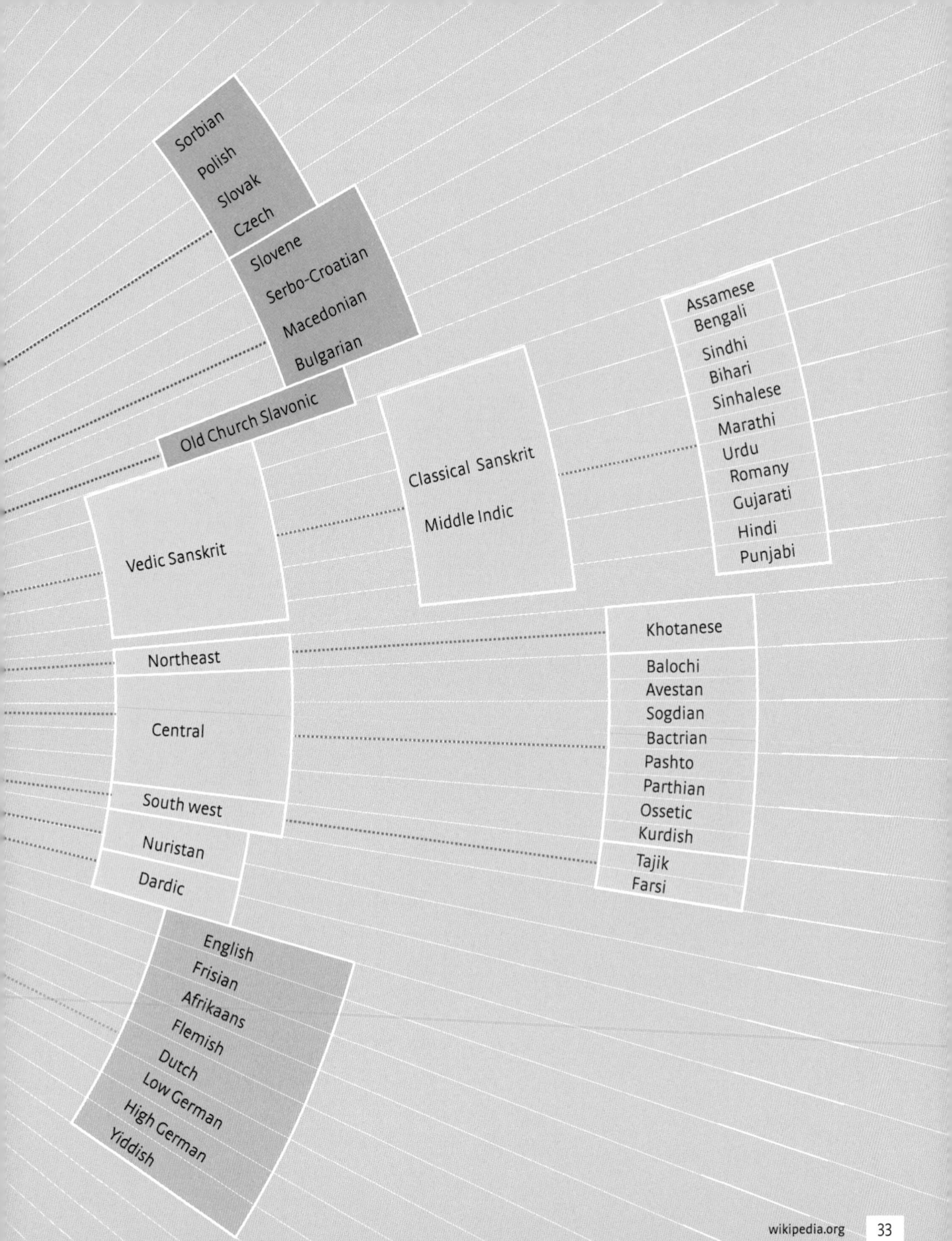
Sorbian
Polish
Slovak
Czech
Slovene
Serbo-Croatian
Macedonian
Bulgarian
Old Church Slavonic
Assamese
Bengali
Sindhi
Bihari
Sinhalese
Marathi
Urdu
Romany
Gujarati
Hindi
Punjabi
Classical Sanskrit
Middle Indic
Vedic Sanskrit
Khotanese
Balochi
Avestan
Sogdian
Bactrian
Pashto
Parthian
Ossetic
Kurdish
Tajik
Farsi
Northeast
Central
South west
Nuristan
Dardic
English
Frisian
Afrikaans
Flemish
Dutch
Low German
High German
Yiddish

THE NEW POWER **GENERATION**

Using the power of the sun, wind and water, every nation should be replacing fossil-fuelled power stations with a greener alternative. These are the most successful Green power-producing countries in the world.

SOLAR POWER AROUND THE WORLD

top ten countries, installed photovoltaic (PV) capacity, 2012, mega watts (MW)

MW	Country
7,630	Germany
4,950	China
4,223	Italy
3,282	USA
1,790	Japan
1,427	India
1,012	France
993	Australia
933	Bulgaria
817	UK

Total worldwide 32,340MW
NB: Solar photovoltaics (PV) directly convert solar energy into electricity.

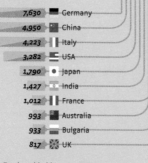

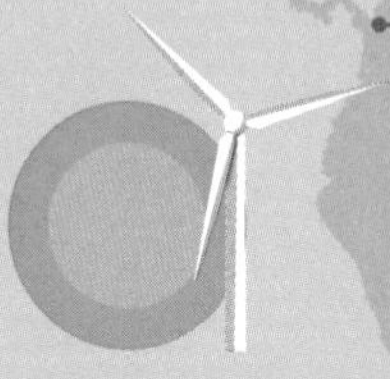

WORLD WIND ENERGY

(2013, by installed capacity, MW)

MW	Country
80,827	China
60,009	USA
30,442	Germany
22,907	Spain
19,564	India
9,610	UK
8,415	Italy
7,821	France
6,578	Canada
4,578	Denmark

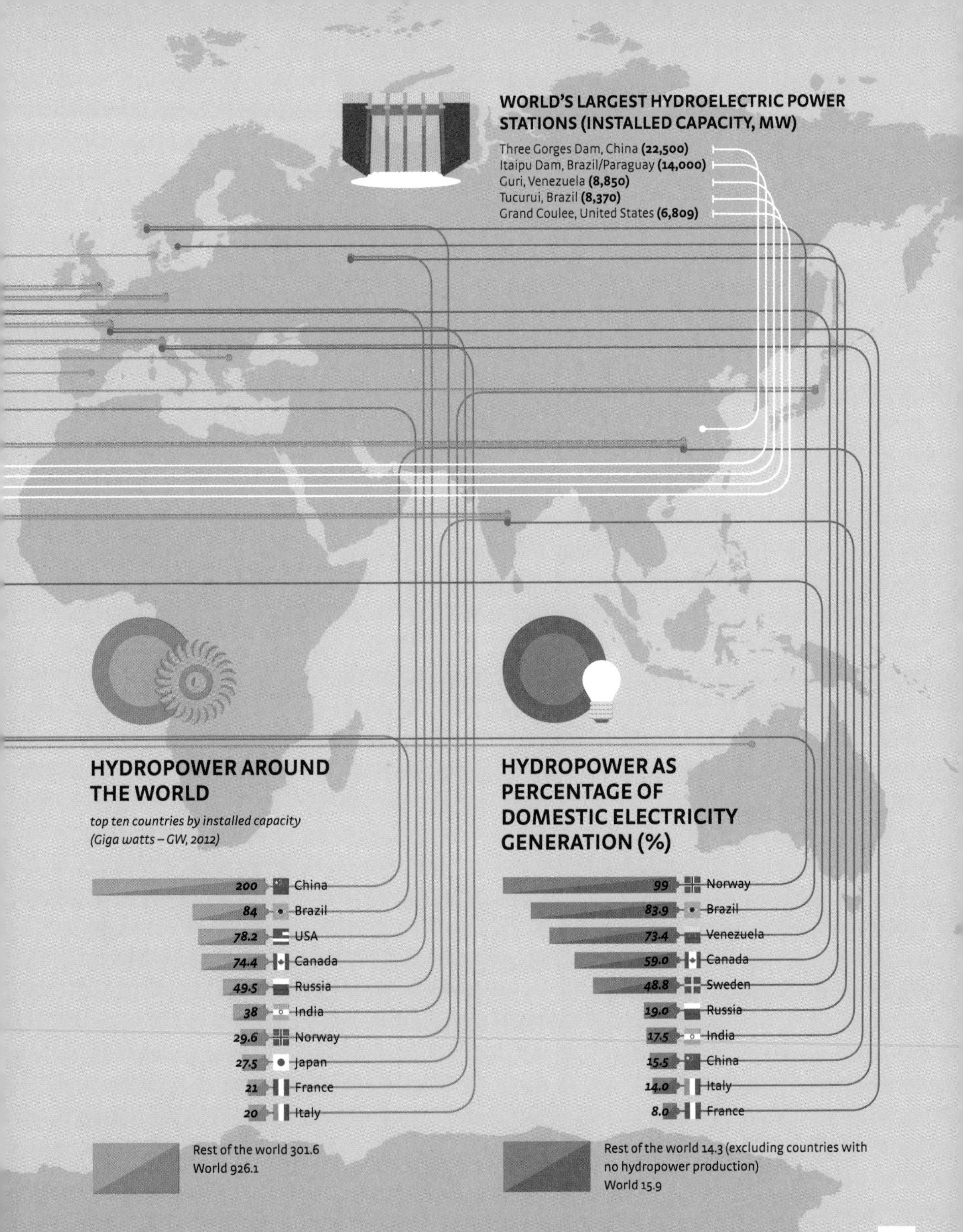
WORLD'S LARGEST HYDROELECTRIC POWER STATIONS (INSTALLED CAPACITY, MW)
Three Gorges Dam, China (22,500)
Itaipu Dam, Brazil/Paraguay (14,000)
Guri, Venezuela (8,850)
Tucurui, Brazil (8,370)
Grand Coulee, United States (6,809)
HYDROPOWER AROUND THE WORLD
top ten countries by installed capacity (Giga watts – GW, 2012)
200 China
84 Brazil
78.2 USA
74.4 Canada
49.5 Russia
38 India
29.6 Norway
27.5 Japan
21 France
20 Italy
Rest of the world 301.6
World 926.1
HYDROPOWER AS PERCENTAGE OF DOMESTIC ELECTRICITY GENERATION (%)
99 Norway
83.9 Brazil
73.4 Venezuela
59.0 Canada
48.8 Sweden
19.0 Russia
17.5 India
15.5 China
14.0 Italy
8.0 France
Rest of the world 14.3 (excluding countries with no hydropower production)
World 15.9

reneweconomy.com.au, wwindea.org, srren.ipcc-wg3.de, earth-policy.org

WORTH MORE **DEAD THAN ALIVE**

The business of organ donation is expensive and sometimes murky. The costs of operations in the USA and other countries are prohibitive to anyone without adequate insurance, which has led to an increase in the black market for human body parts. If you wanted to sell a bit of yourself, here's a guide to what things cost.

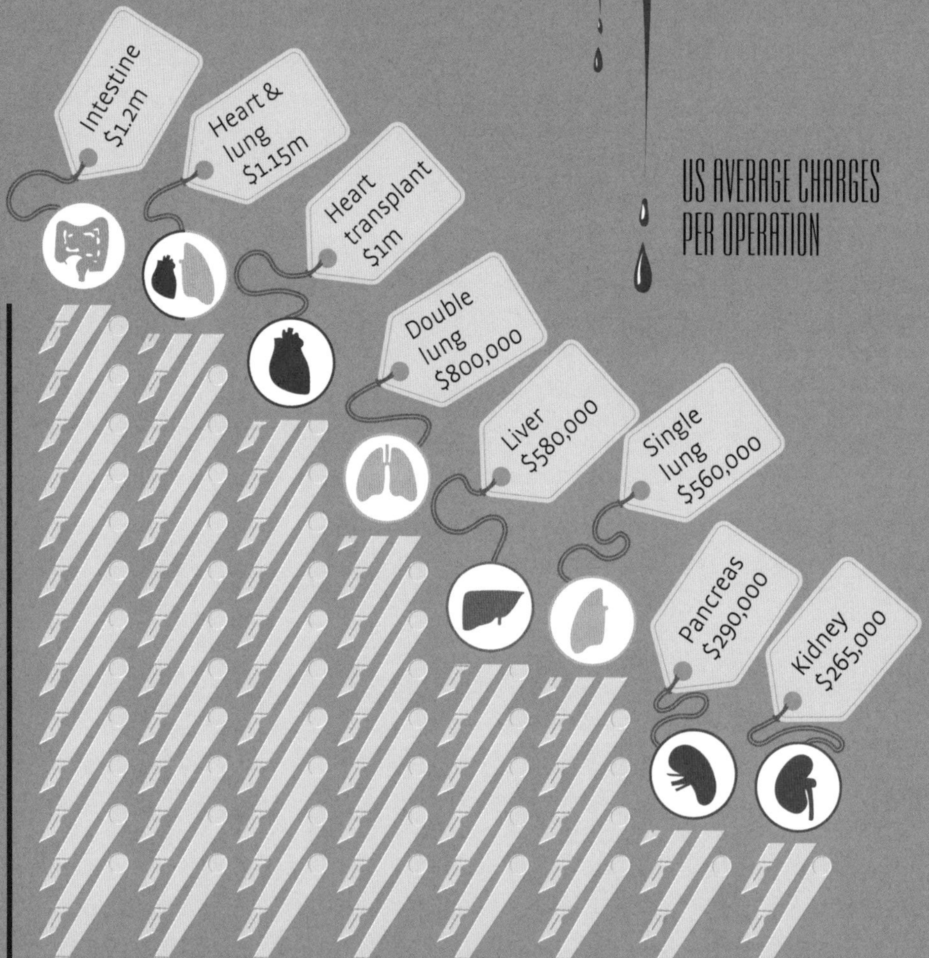

US BLACK MARKET PRICES TO BUY (OTHER TERRITORIES)

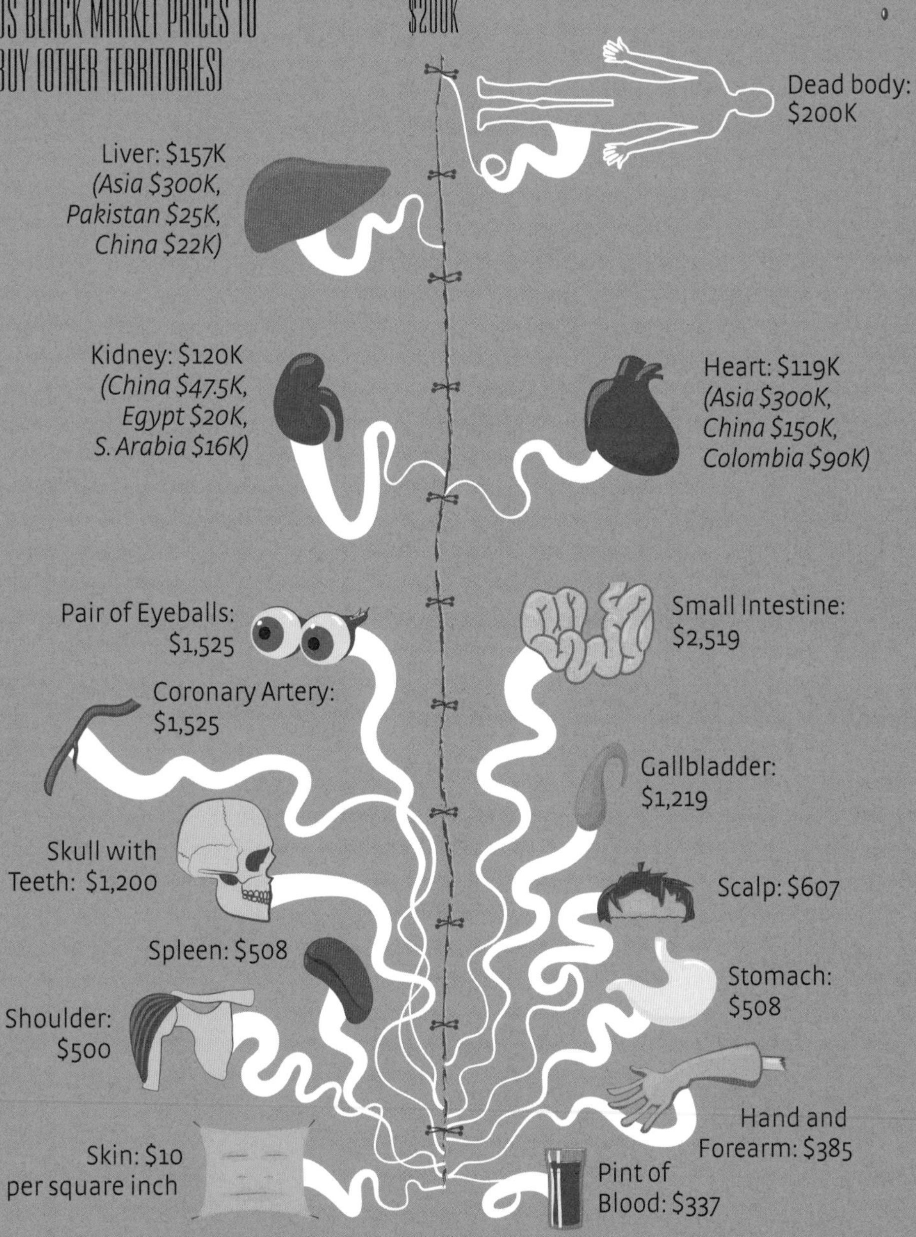

117K bodies reported stolen in the US in one year

gizmodo.com, statisticbrain.com, guardian.co.uk, havocscope.com

THE SUN HAS GOT **HIS HEAT ON**

The Sun makes up over 99.8% of mass of the entire solar system. It is 743 times the total mass of all the planets in the solar system combined and without it we'd be dead. With it the earth has to be just the right distance away or it'd burn up.

1.3
MILLION
EARTHS
CAN FIT
INSIDE
THE
SUN

Mass:

1,989,100,000,000,000, 000,000,000,000,000 kg / 4,385,215,000,000,000, 000,000,000,000,000lbs

Temperature:

Core of the Sun is around

15,000,000° C / 27,000,032°F

The temperature of the surface (photosphere) is about 5,500°C/9,932°F

Element	% of total atoms	% of total mass
Hydrogen	91.2	71.0
Helium	8.7	27.1
Oxygen	0.078	0.97
Carbon	0.043	0.40
Nitrogen	0.0088	0.096
Silicon	0.0045	0.099
Magnesium	0.0038	0.076
Neon	0.0035	0.058
Iron	0.030	0.014
Sulfur	0.015	0.040

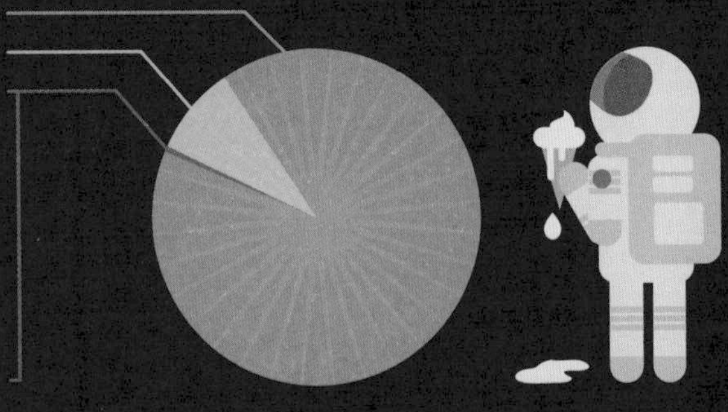

Encyclopedia Britannica, solarsystem.nasa.gov

THE **47–HOUR** DAY

If you board a plane just to the west of the International Date Line (IDL) at midnight and fly west, as the Earth turns below you'll be moving with the sun and 'slowing' time. Don't fly over more than one time zone per hour and you can stretch your day to a theoretical maximum of 47 hrs 59 mins 59 secs.

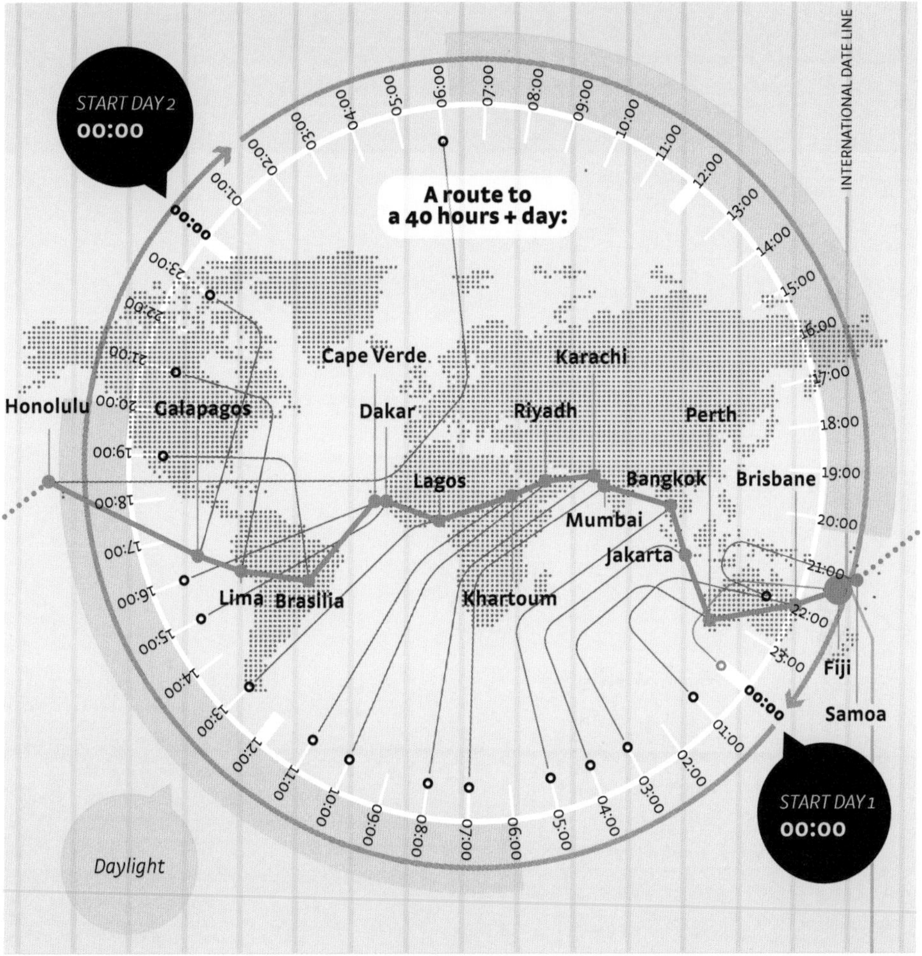

Or

1. You may cross the IDL eastwards to subtract 24 hours.
2. Thus the date to the west of the line is repeated after the following midnight
3. Spend a day in Hong Kong and fly to Honolulu to enjoy the 'same' day again

NEW PLANET, **NEW YOU**

How to lose weight and look tanned without all that gym and solarium work? Move to another planet. Here's a handy planet primer that shows how close to the sun and how weightless you'll feel on each.

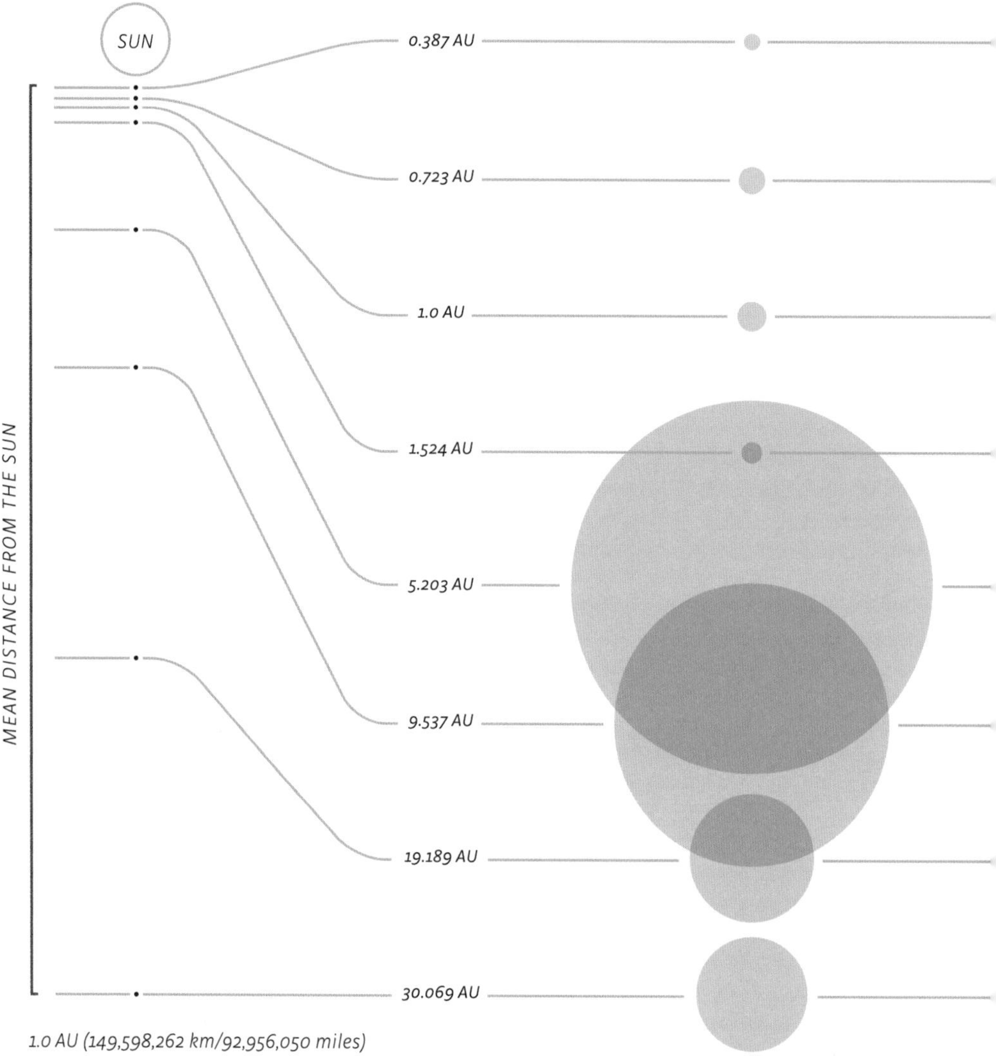

1.0 AU (149,598,262 km/92,956,050 miles)

	Gravity	Length of day Earth = 1	Average Surface Temperature	Moons	Average Orbit Speed (km/h)	Diameter (km/miles)	Atmosphere
Mercury	22kg/49lb	0.67	167C° 333F°	none	173,326	4878/ 3031	oxygen sodium hydrogen helium potassium
Venus	54kg/119lb	243.5	464C° 867F°	none	170,503	12,092/ 7514	carbon dioxide nitrogen sulphuric acid
Earth	60kg/132lb	1	15C° 59F°	1	107,218	12,742/ 7918	nitrogen oxygen other gases
Mars	22kg/49lb	1.026	-165C° -265F°	2	86,677	6792/ 4220	carbon dioxide nitrogen argon
Jupiter	28kg/62lb	0.41	-110C° -166F°	67	47,002	142,984/ 88,846	hydrogen helium
Saturn	63kg/139lb	0.44	-140C° -220F°	62	34,701	120,536/ 74,898	hydrogen helium
Uranus	51kg/112lb	0.72	-195C° -319F°	27	24,477	51,118/ 31,763	hydrogen helium methane
Neptune	66kg/145lb	0.67	-200C° -328F°	14	19,566	49,532/ 30,778	hydrogen helium methane

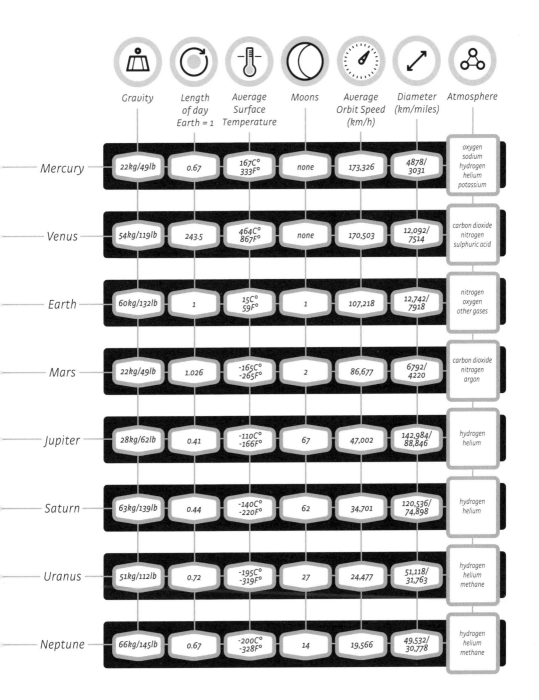

CIRCLES **OF LIFE**

We're always being told how the future population of the world will change radically from what it is now, but really, will it? Here are the projected estimates of how the gender and age ranges will differ globally over the next century.

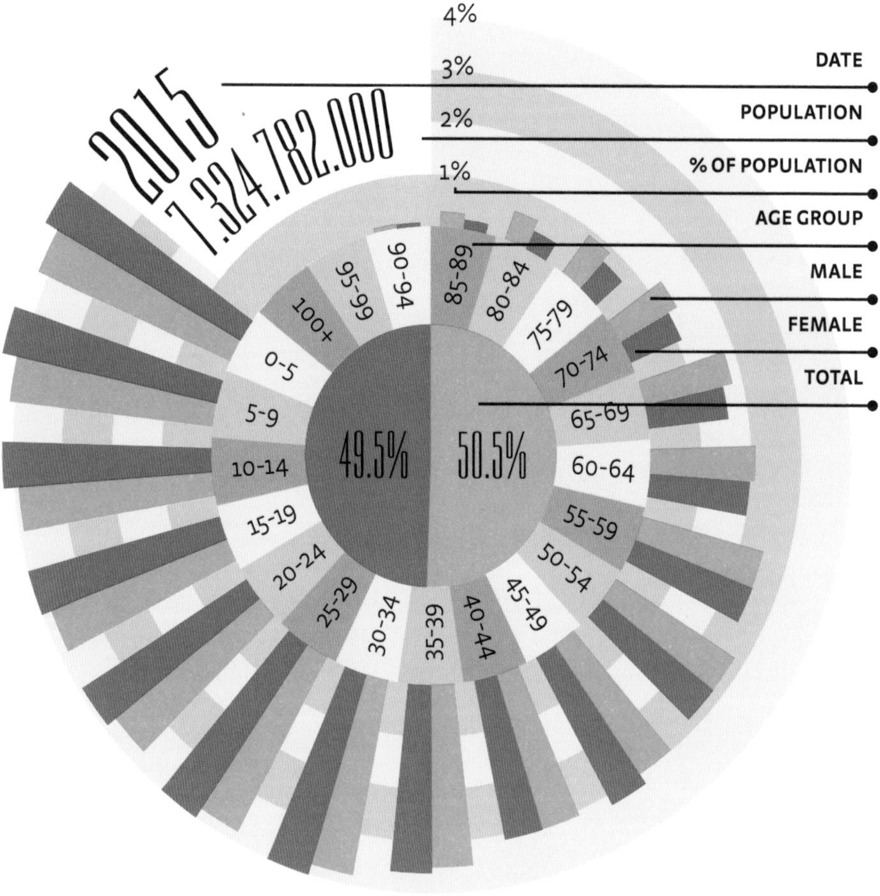

Note: *The total population doesn't add up to a perfect 100% due to a round down of the centesimal cases of the data. This means the data isn't 100% because it has been previously manipulated by probability models and software that didn't include the centesimal case!*

 populationpyramid.net, un.org, economist.com, esa.un.org

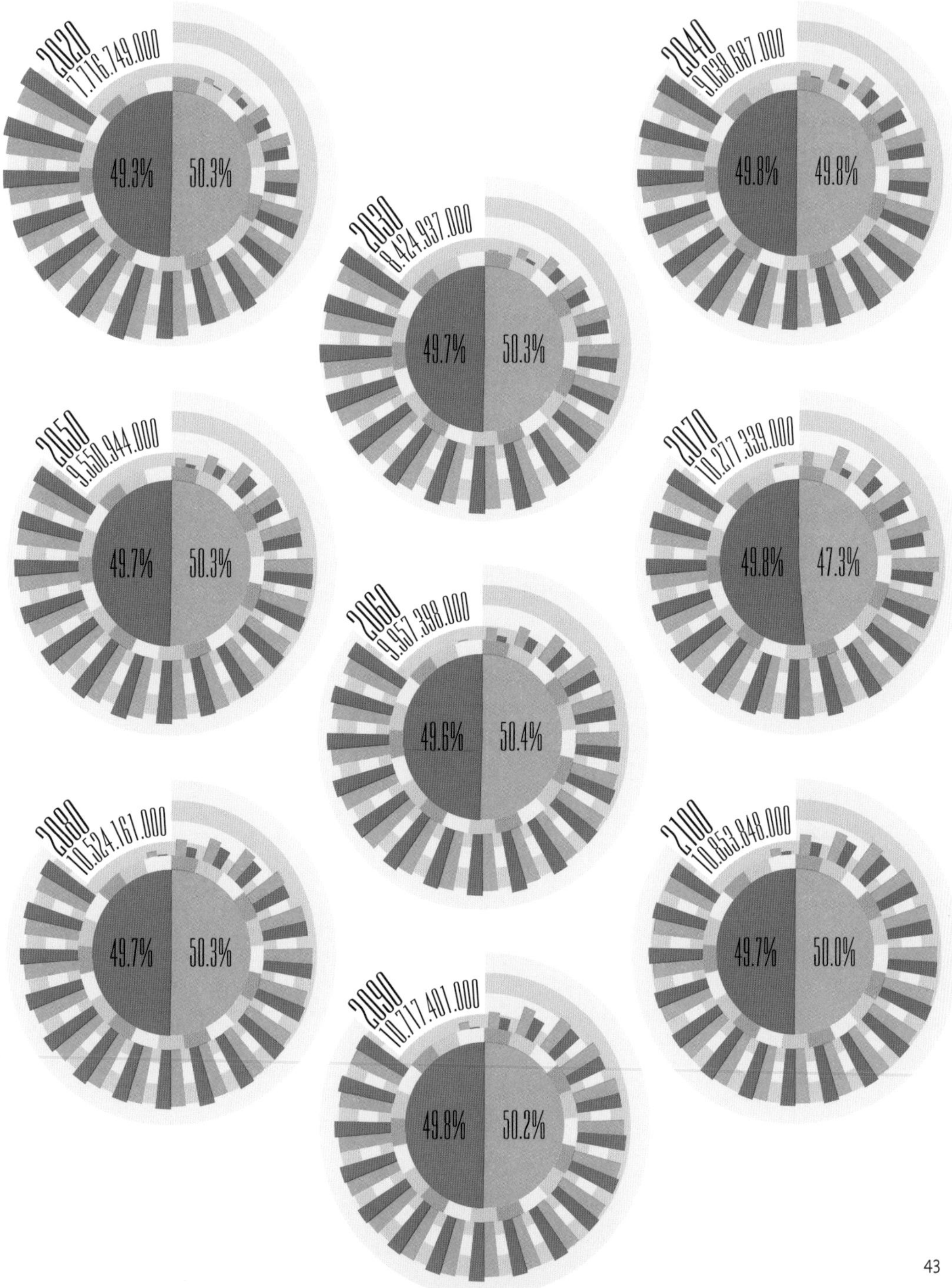
2020
7.716.749.000
49.3%
50.3%
2030
8.424.937.000
49.7%
50.3%
2040
9.038.687.000
49.8%
49.8%
2050
9.550.944.000
49.7%
50.3%
2060
9.957.398.000
49.6%
50.4%
2070
10.277.339.000
49.8%
47.3%
2080
10.524.161.000
49.7%
50.3%
2090
10.717.401.000
49.8%
50.2%
2100
10.853.848.000
49.7%
50.0%

TINY **NOISEMAKERS**

Here's how the lesser water boatman insect is louder than an elephant, although smaller than a common fruit fly – because it's the loudest creature ever recorded, relatively speaking. The units shown are a ratio of decibels after the acoustic pressure has been adjusted for body size.

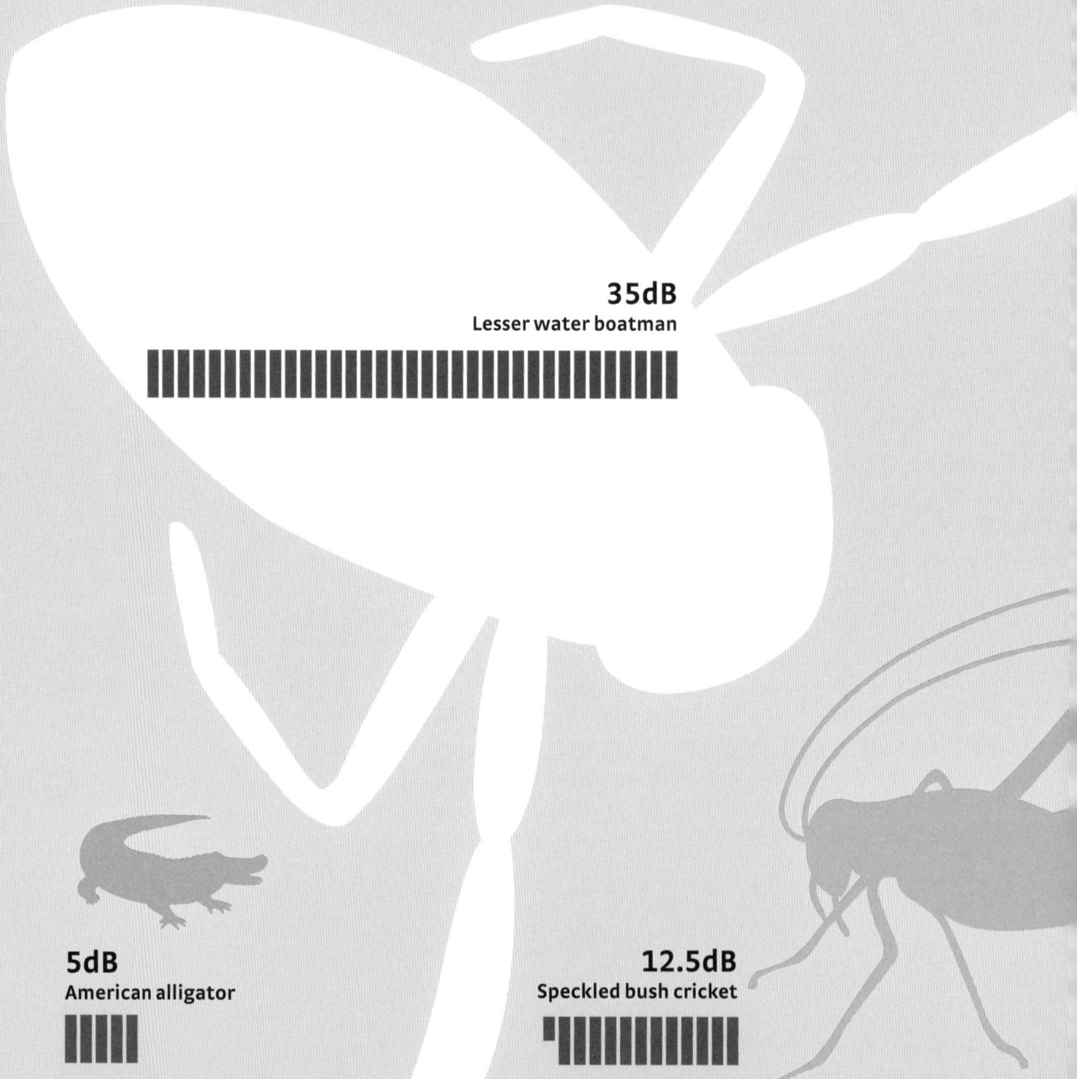

 Dr James Windmill, University of Strathclyde

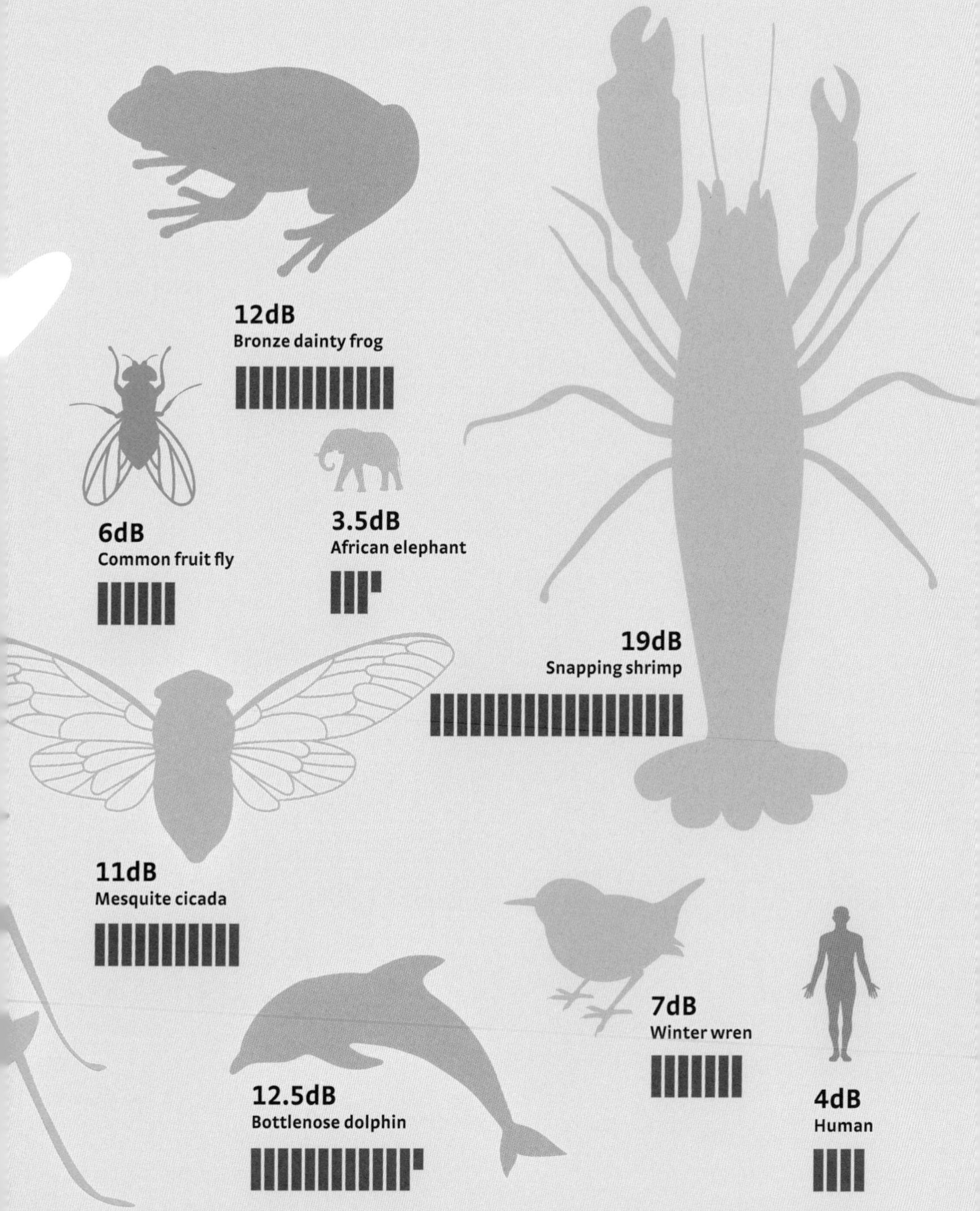
12dB
Bronze dainty frog
6dB
Common fruit fly
3.5dB
African elephant
19dB
Snapping shrimp
11dB
Mesquite cicada
7dB
Winter wren
12.5dB
Bottlenose dolphin
4dB
Human

UP, UP **AND AWAY**

Air travel hubs that handle the most traffic are not always filled with passengers, as this comparison between the busiest airports for planes landing and taking off, and the busiest by numbers of passengers, demonstrates.

TOP 10 AIRPORTS
by *total aircraft landings* and *take-offs* per annum:

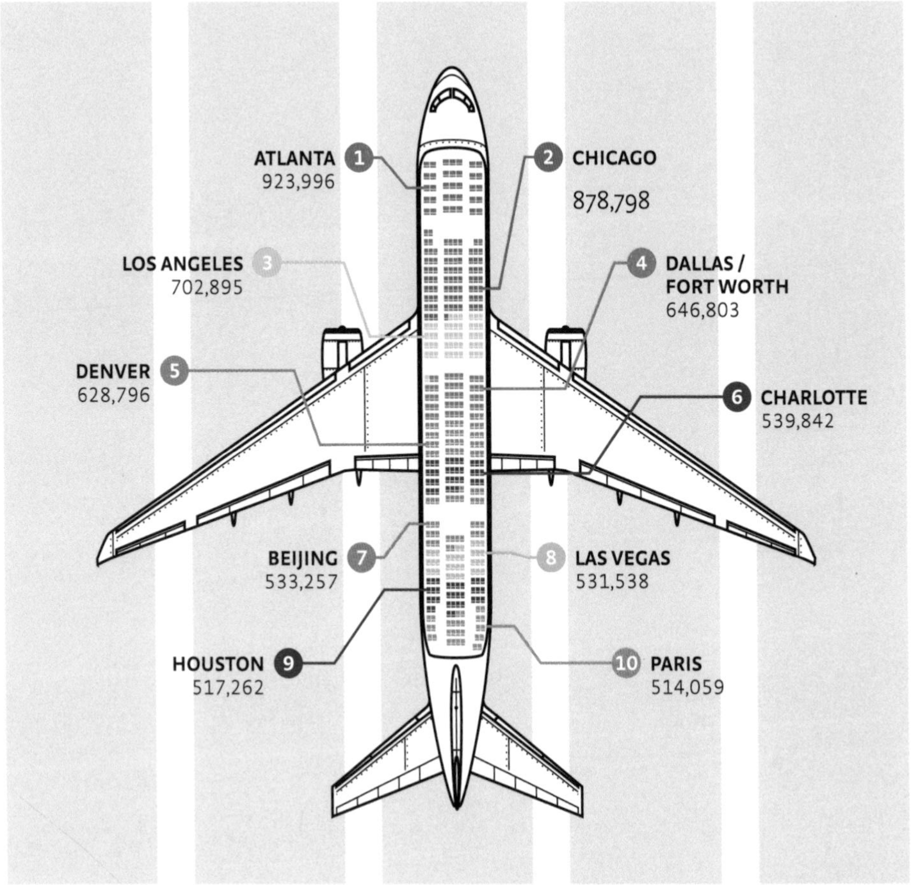

Total number of Planes in Top 10: 6,417,246
Total number of Seats on Plane: 425

1 x Seat = 15,060 Planes

TOP 10 AIRPORTS
by *passenger volume* per annum:

Total number of passengers in Top 10: 660,213,679
Total number of seats on Plane: 425

1 x Seat = 1,553,444 Passengers

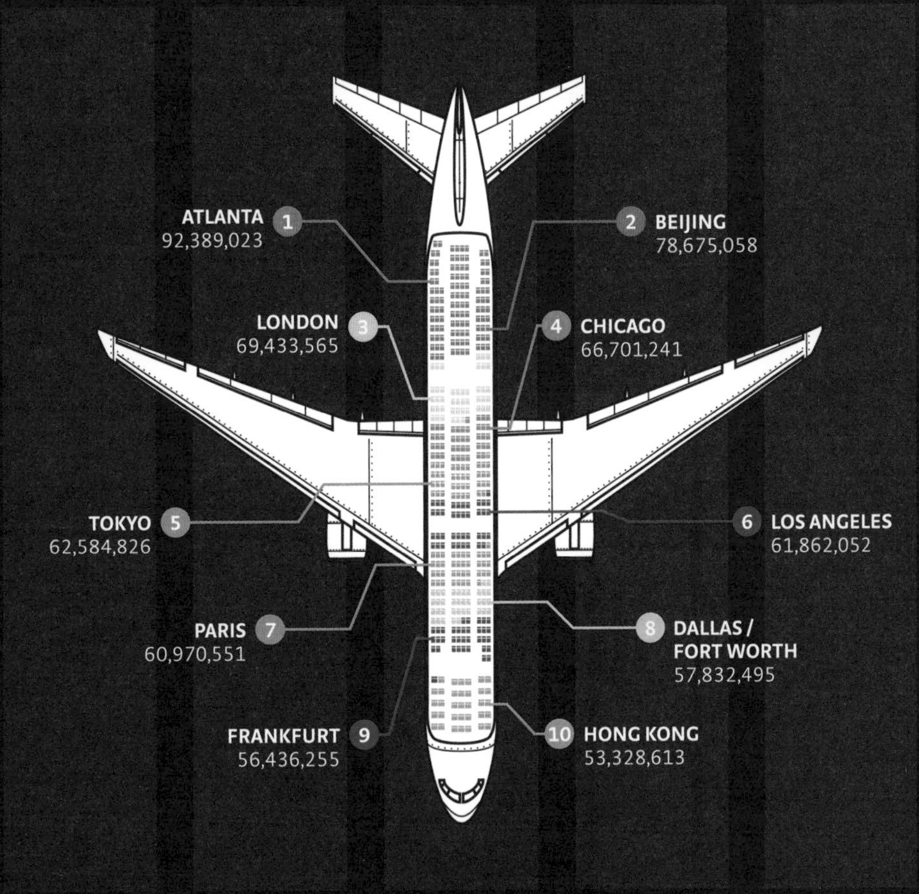

It is estimated 400,000 people are in flight at any one time around the world.

In the USA 15,500 air traffic controllers handle 50,000 flights per day, and each day 1.7 million passengers board a plane in the US
= 1 air traffic controller handles approx. 110 people in the air at any one time

aci.aero/data-centre/annual-traffic-data, theatlantic.com/technology, faa.gov, natca.org, airport-technology.com

HIGH ANXIETY

Nicknamed 'Spiderman', Frenchman Alain Robert is an urban climber who has scaled many of the world's tallest structures. He can take up to six hours to ascend a building, but a fall would take only seconds. Here's exactly how many seconds for the buildings he has already climbed, plus a few that he might consider.

HEIGHT **381m/**1250ft

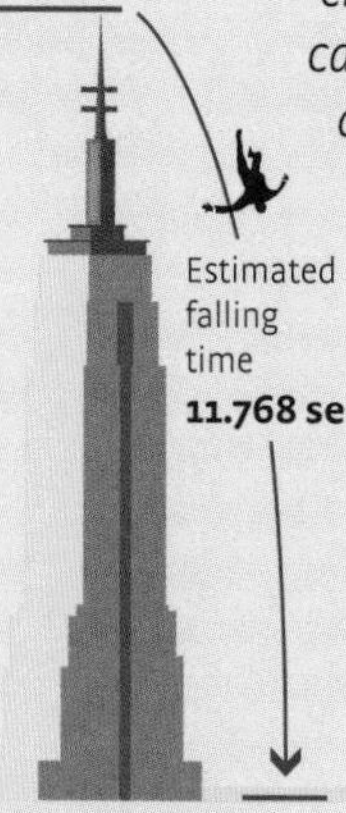

Estimated falling time **11.768 sec**

CLIMBED 1994

EMPIRE STATE BUILDING, NYC, USA

Falling speed **43.37 m/ps** (142.29 ft/ps)

HEIGHT **106m/**347.77ft

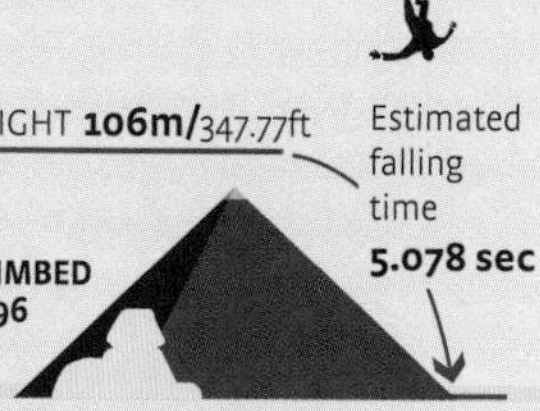

Estimated falling time **5.078 sec**

CLIMBED 1996

LUXOR HOTEL, LAS VEGAS, USA

Falling speed **35.63 m/ps** (116.89 ft/ps)

HEIGHT **227m/**744.75ft

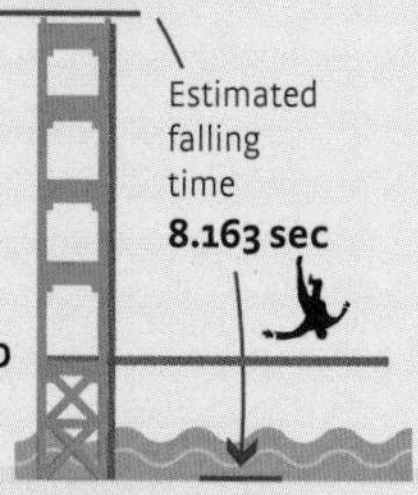

Estimated falling time **8.163 sec**

CLIMBED 1996

GOLDEN GATE BRIDGE, SAN FRANCISCO, USA

Falling speed **41.61 m/ps** (136.51 ft/ps)

HEIGHT **442m/**1450.13ft

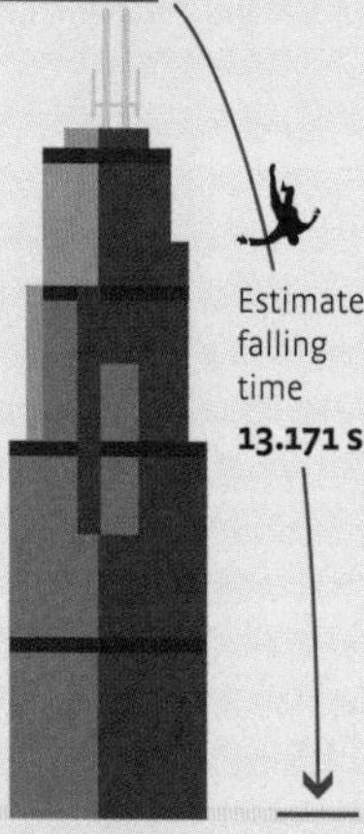

Estimated falling time **13.171 sec**

CLIMBED 1999

WILLIS TOWER, CHICAGO, USA

Falling speed **43.58 m/ps** (142.98 ft/ps)

HEIGHT **235m/**770.997ft

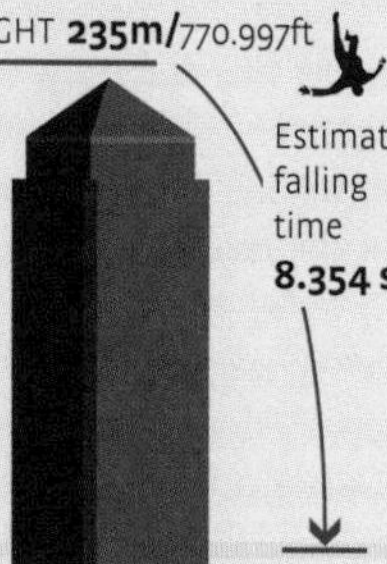

Estimated falling time **8.354 sec**

CLIMBED 2002

ONE CANADA SQUARE, LONDON, ENGLAND

Falling speed **41.78 m/ps** (137.07 ft/ps)

HEIGHT **509m/**1669.95ft

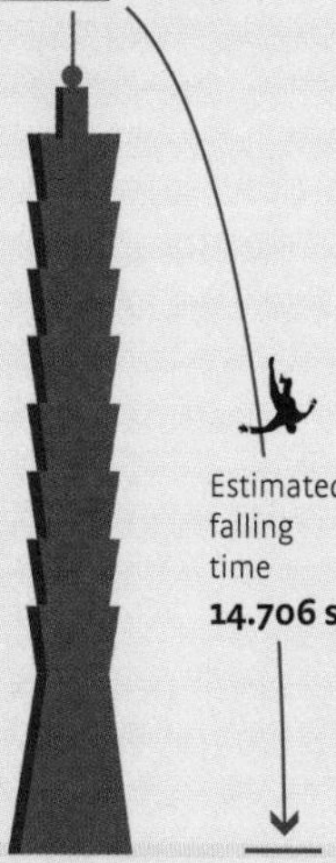

Estimated falling time **14.706 sec**

CLIMBED 2004

TAIPEI 101, TAIPEI, REPUBLIC OF CHINA (TAIWAN)

Falling speed **43.70 m/ps** (143.37 ft/ps)

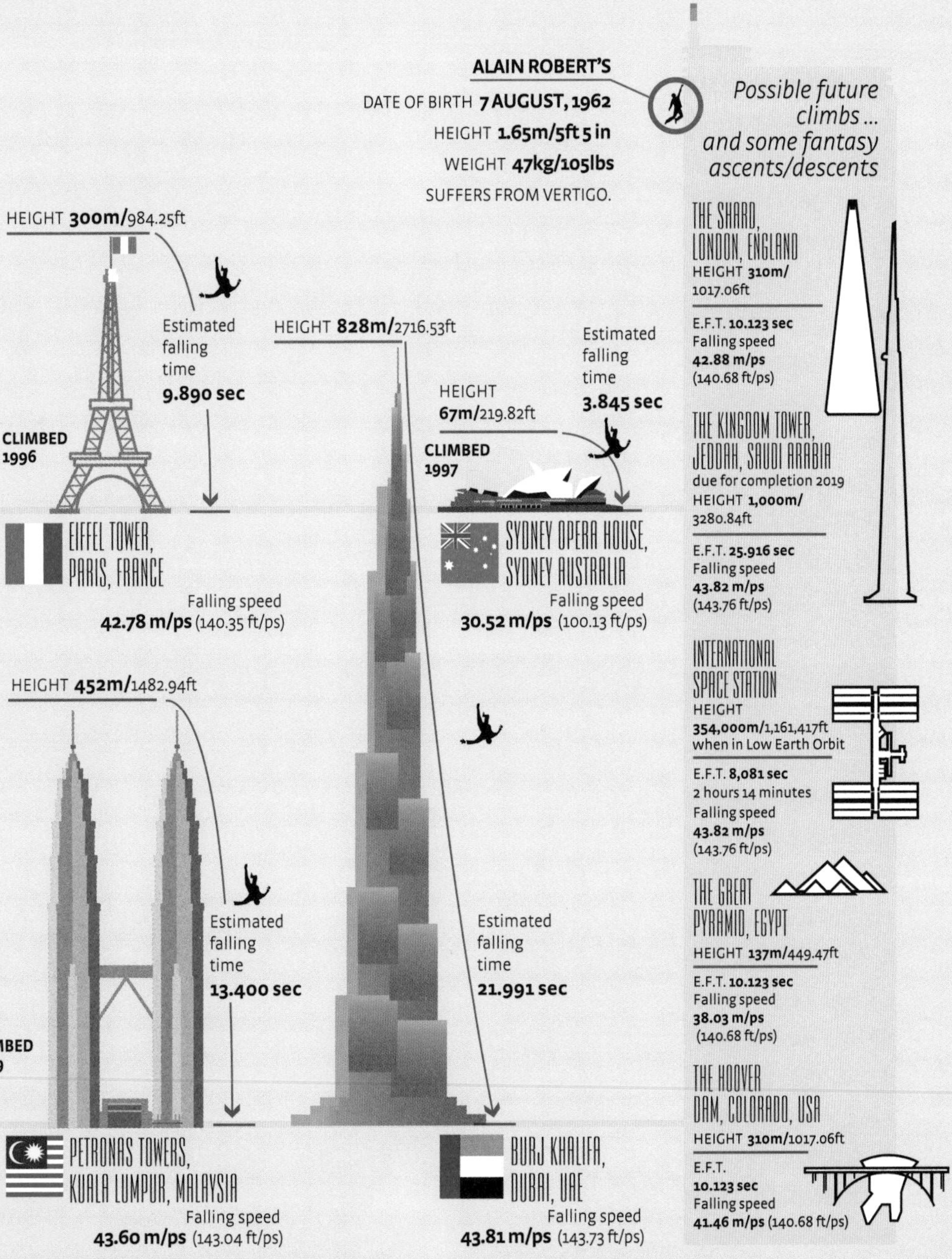
ALAIN ROBERT'S
DATE OF BIRTH 7 AUGUST, 1962
HEIGHT 1.65m/5ft 5 in
WEIGHT 47kg/105lbs
SUFFERS FROM VERTIGO.
HEIGHT 300m/984.25ft
Estimated falling time 9.890 sec
CLIMBED 1996
EIFFEL TOWER, PARIS, FRANCE
Falling speed 42.78 m/ps (140.35 ft/ps)
HEIGHT 828m/2716.53ft
HEIGHT 67m/219.82ft
Estimated falling time 3.845 sec
CLIMBED 1997
SYDNEY OPERA HOUSE, SYDNEY AUSTRALIA
Falling speed 30.52 m/ps (100.13 ft/ps)
HEIGHT 452m/1482.94ft
Estimated falling time 13.400 sec
CLIMBED 2009
PETRONAS TOWERS, KUALA LUMPUR, MALAYSIA
Falling speed 43.60 m/ps (143.04 ft/ps)
Estimated falling time 21.991 sec
BURJ KHALIFA, DUBAI, UAE
Falling speed 43.81 m/ps (143.73 ft/ps)
Possible future climbs ... and some fantasy ascents/descents
THE SHARD, LONDON, ENGLAND
HEIGHT 310m/1017.06ft
E.F.T. 10.123 sec
Falling speed 42.88 m/ps (140.68 ft/ps)
THE KINGDOM TOWER, JEDDAH, SAUDI ARABIA
due for completion 2019
HEIGHT 1,000m/3280.84ft
E.F.T. 25.916 sec
Falling speed 43.82 m/ps (143.76 ft/ps)
INTERNATIONAL SPACE STATION
HEIGHT 354,000m/1,161,417ft when in Low Earth Orbit
E.F.T. 8,081 sec
2 hours 14 minutes
Falling speed 43.82 m/ps (143.76 ft/ps)
THE GREAT PYRAMID, EGYPT
HEIGHT 137m/449.47ft
E.F.T. 10.123 sec
Falling speed 38.03 m/ps (140.68 ft/ps)
THE HOOVER DAM, COLORADO, USA
HEIGHT 310m/1017.06ft
E.F.T. 10.123 sec
Falling speed 41.46 m/ps (140.68 ft/ps)

alainrobert.com, keisan.casio.com, wikipedia.org, nasa.gov, pbs.org, usbr.gov

CROSSTOWN TRAFFIC **CRAZY**

International traffic congestion in the world's leading cities is always big news, and there are lots of surveys published regularly that tell us where it's worst, the slowest and most annoying. They don't all agree, either.

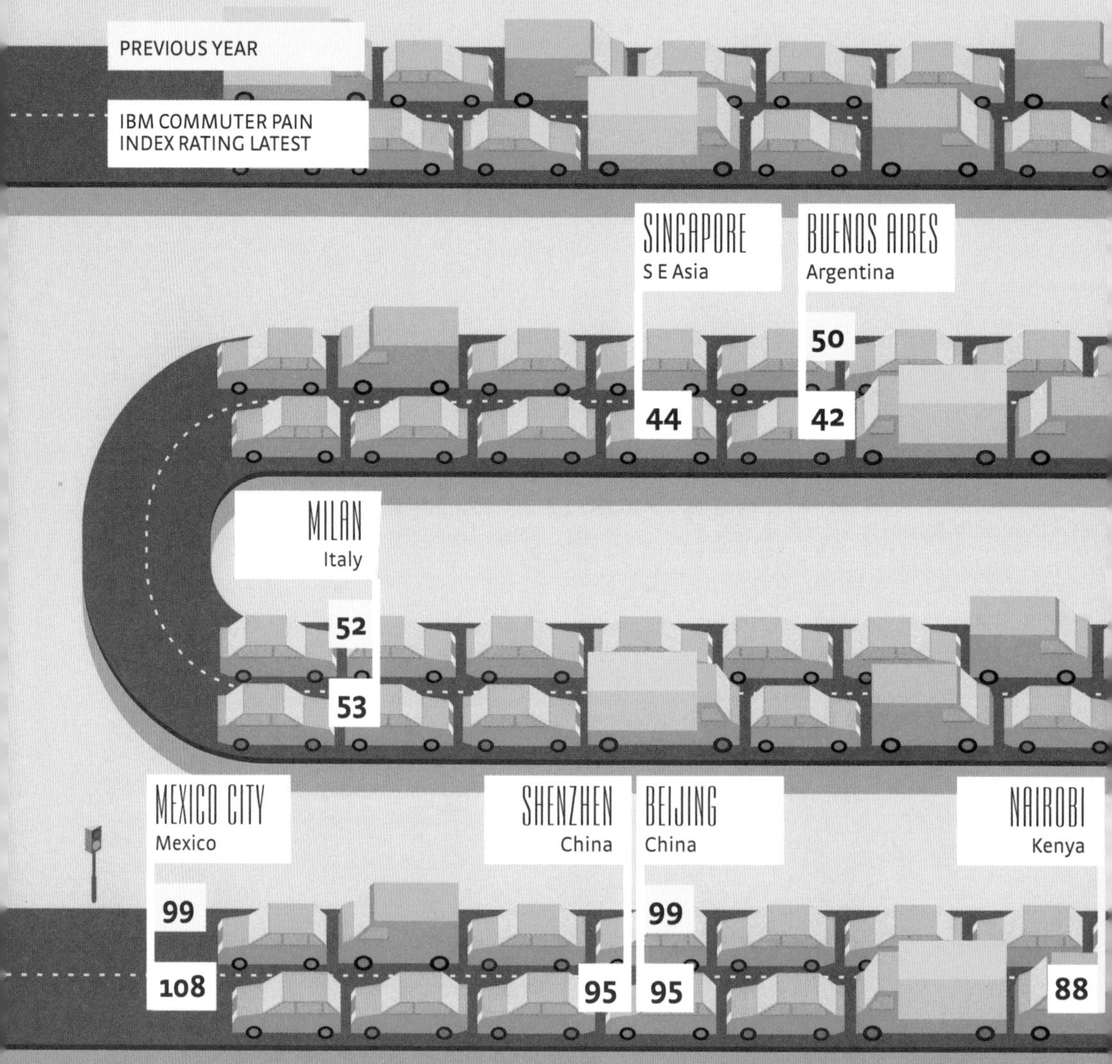

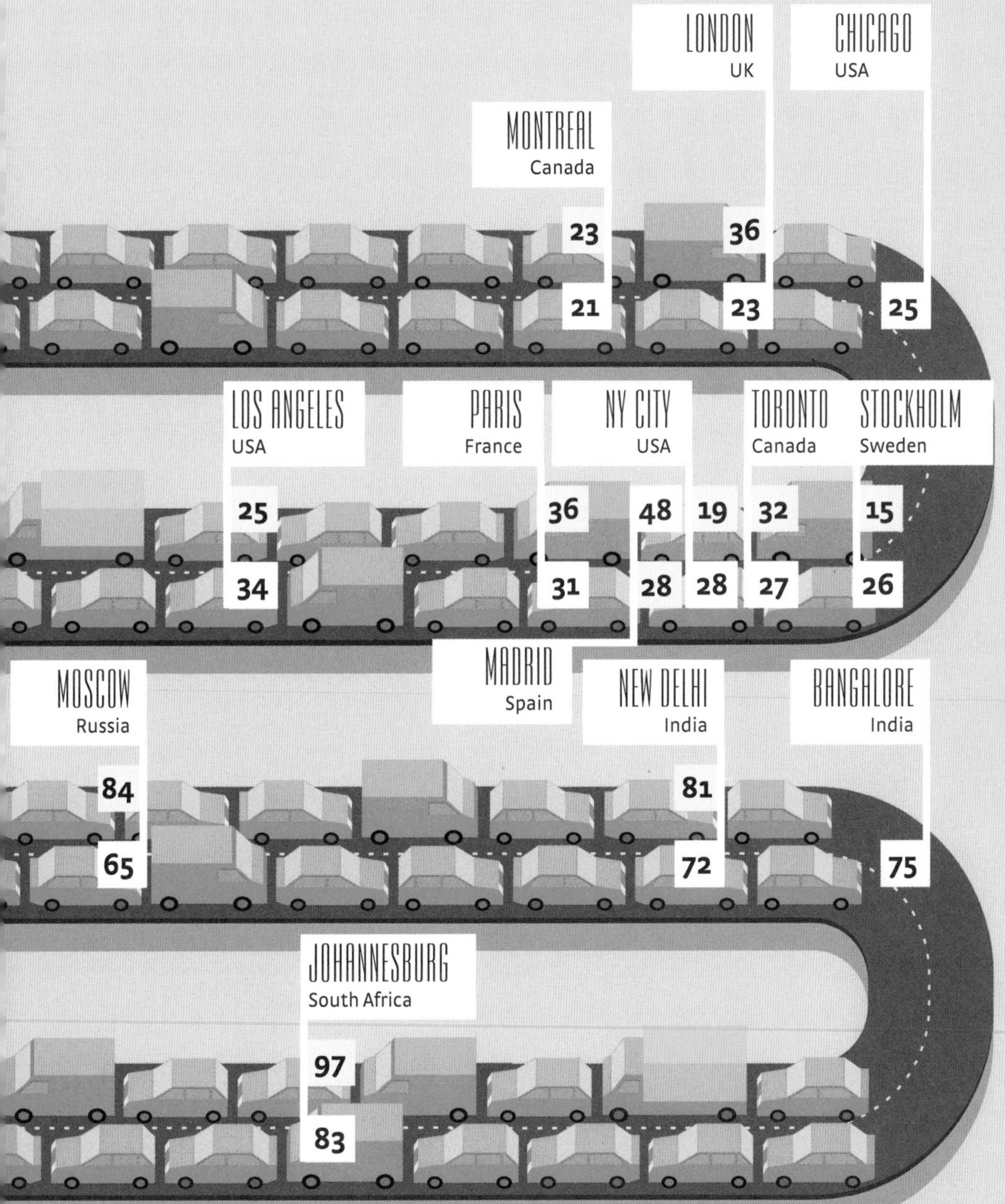

IBM, wikipedia.org, forbes.com

THESE **AREN'T** THE PLANETS YOU ARE LOOKING FOR

Darth Vader boasted in Star Wars *(1977) that, 'the ability to destroy a planet is insignificant next to the power of the Force', and the* Death Star *was roughly a quarter the size of Earth. Here's how much power the Force would expend in destroying planets of our universe. Of course, Obi-Wan could distract it with the explanation that they're not the ones it's looking for...*

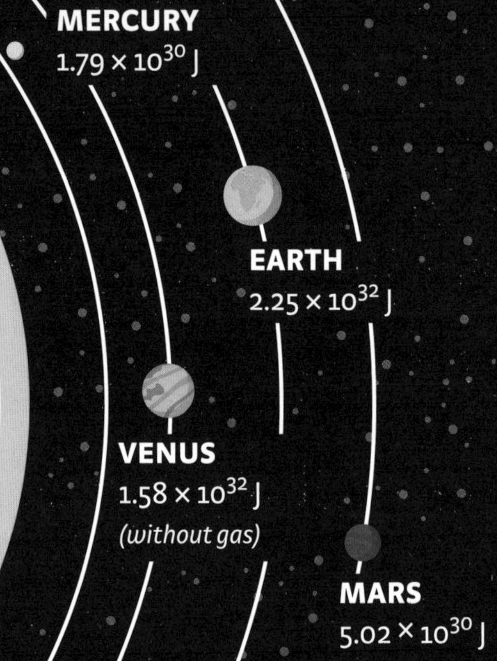

The Equation

$$U = \frac{GM_p^2}{5R_p}$$

G: *Gravitational Constant*

Mp: *Planet Mass*

Rp: *Planet Radius*

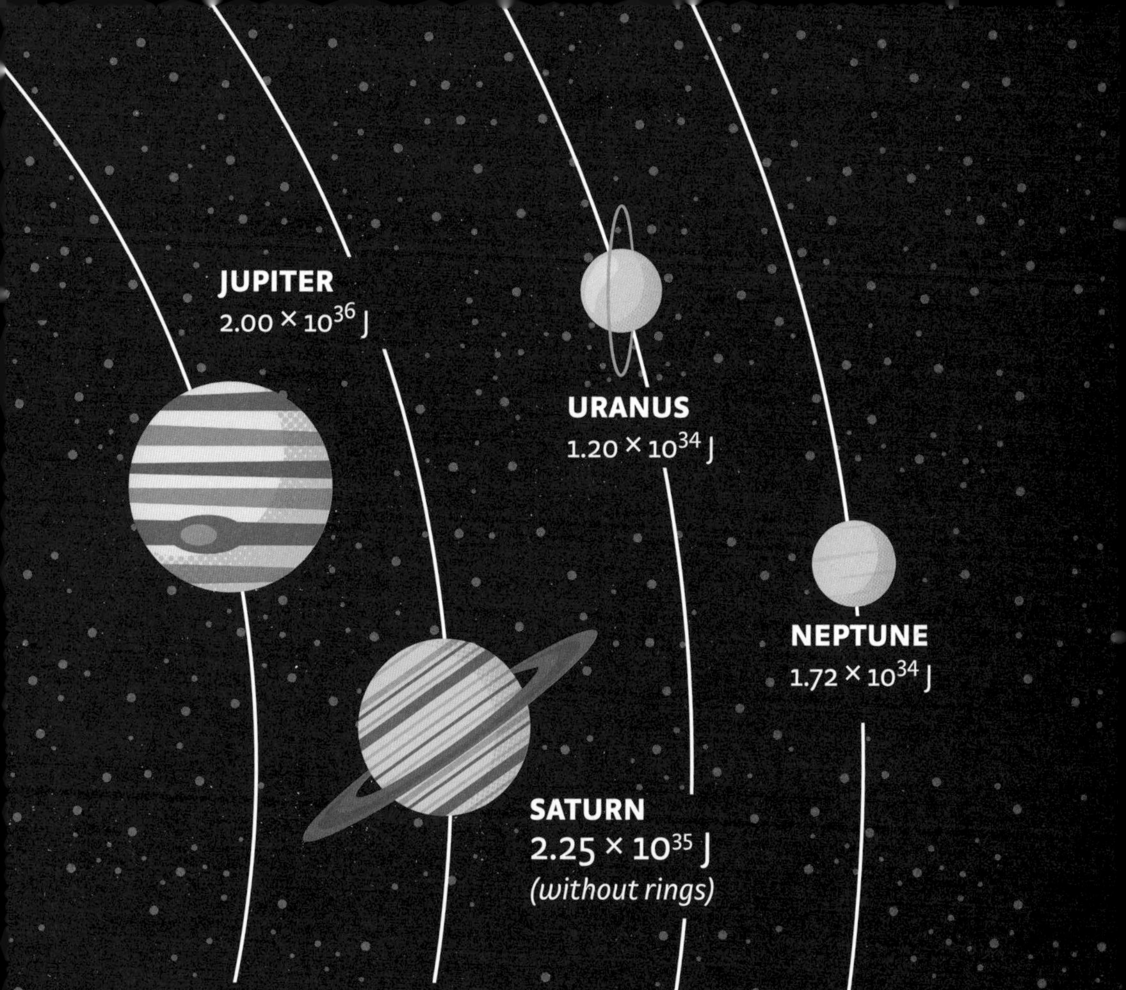
JUPITER
2.00×10^{36} J
URANUS
1.20×10^{34} J
NEPTUNE
1.72×10^{34} J
SATURN
2.25×10^{35} J
(without rings)

SIZE COMPARISON

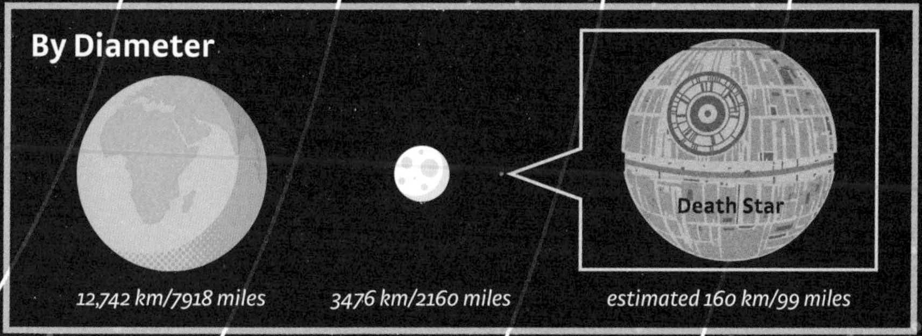
By Diameter
Death Star
12,742 km/7918 miles
3476 km/2160 miles
estimated 160 km/99 miles

MEGACITIES OF THE FUTURE

The UN estimates that by 2030, nearly 60% of us will live in urban areas. Cities occupy just 2% of the Earth's land, but represent most of the global energy consumption. In the coming decades, 95% of urban growth will happen in the developing world, as this chart of city growth from 1950 to 2025 shows.

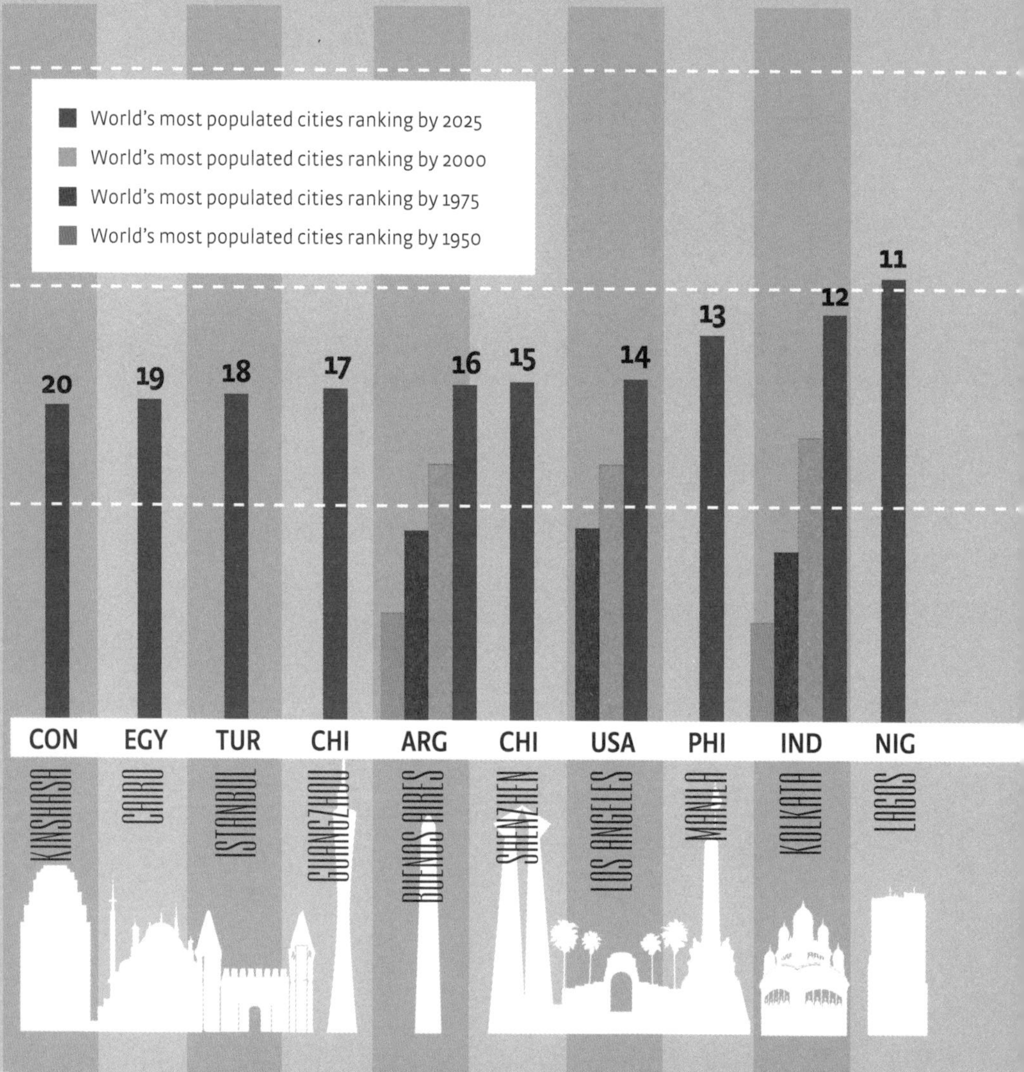

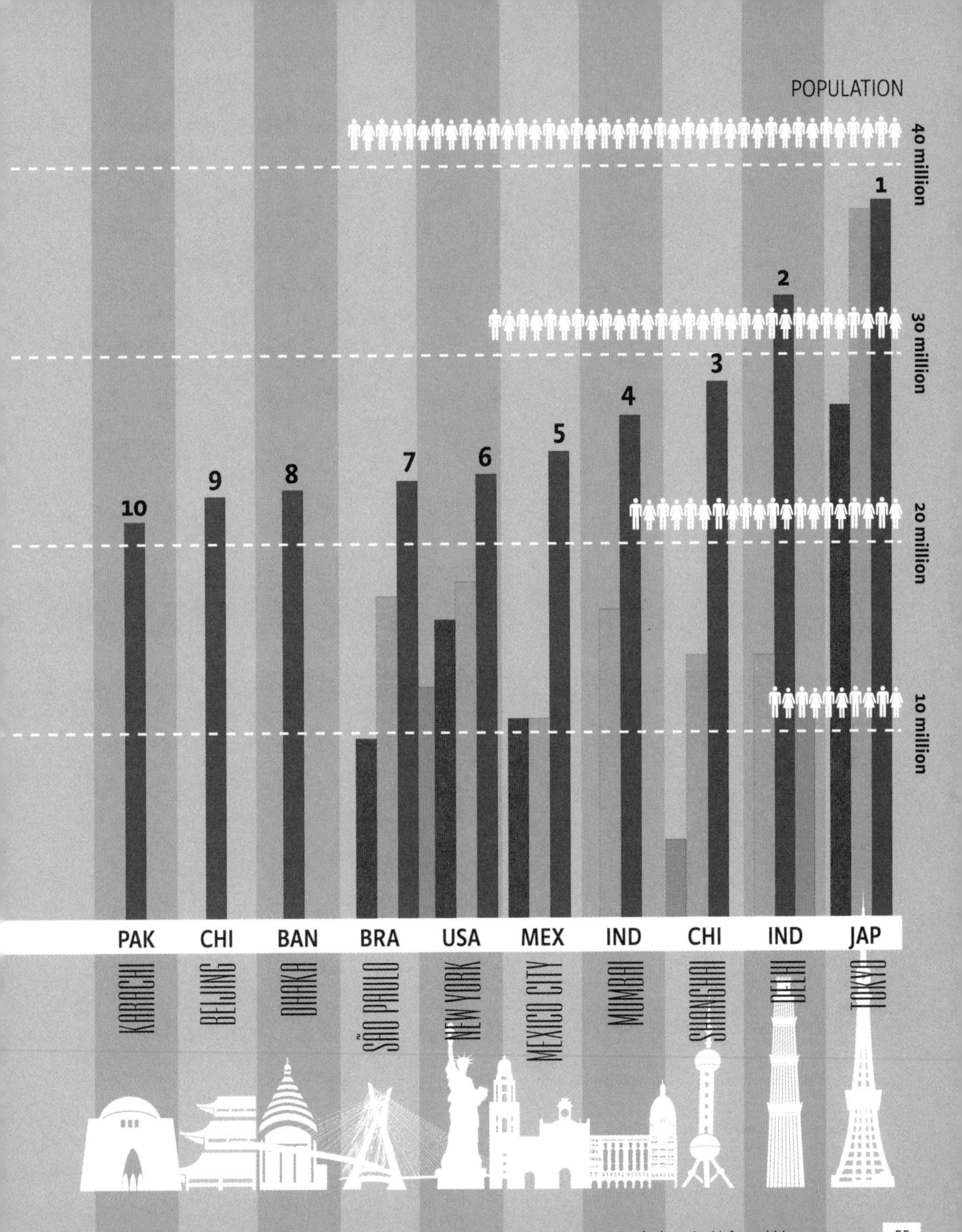
POPULATION
40 million
30 million
20 million
10 million
1
2
3
4
5
6
7
8
9
10
PAK
CHI
BAN
BRA
USA
MEX
IND
CHI
IND
JAP
KARACHI
BEIJING
DHAKA
SÃO PAULO
NEW YORK
MEXICO CITY
MUMBAI
SHANGHAI
DELHI
TOKYO

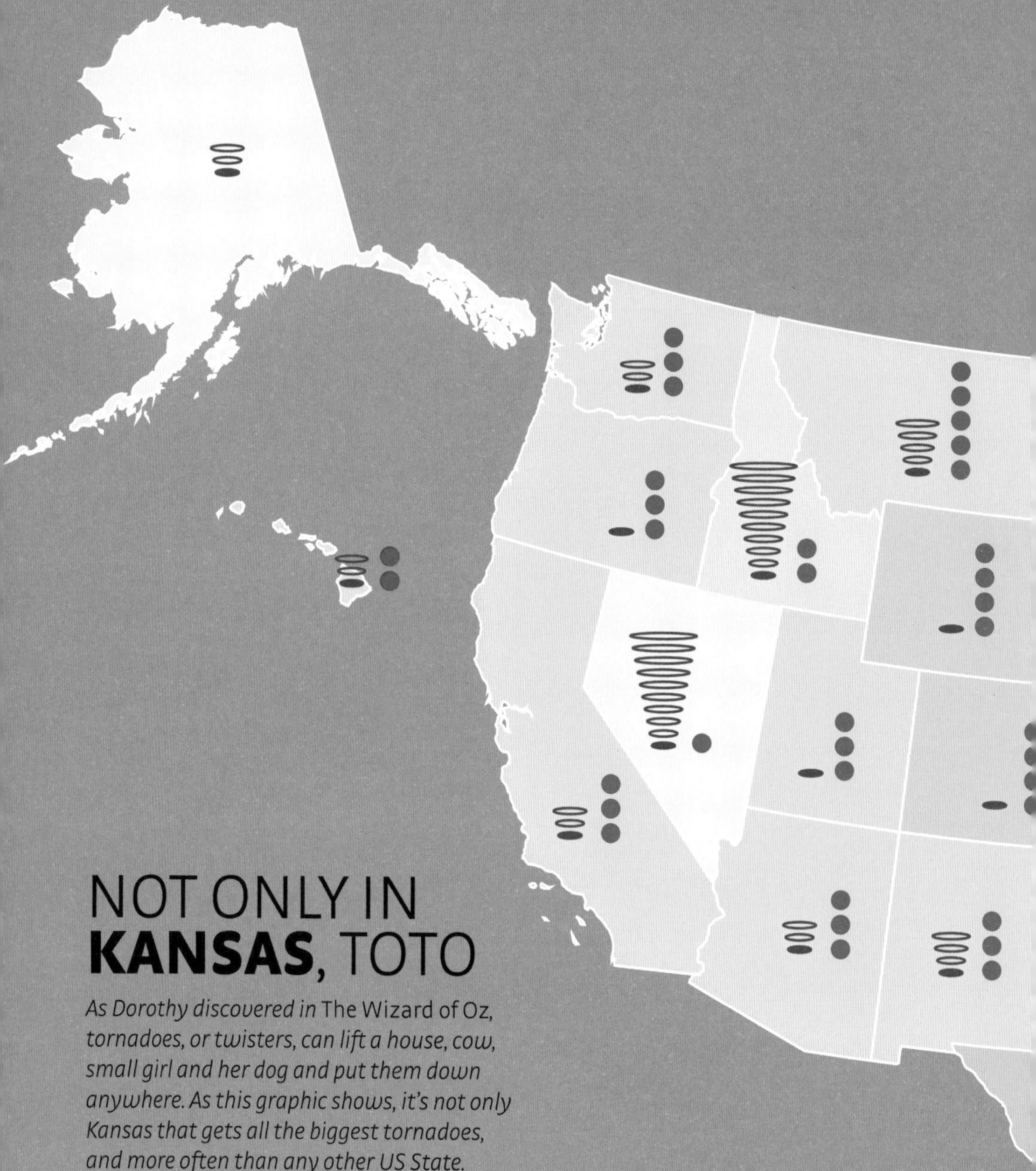

NOT ONLY IN **KANSAS**, TOTO

As Dorothy discovered in The Wizard of Oz, *tornadoes, or twisters, can lift a house, cow, small girl and her dog and put them down anywhere. As this graphic shows, it's not only Kansas that gets all the biggest tornadoes, and more often than any other US State. That dubious honour goes to Arkansas.*

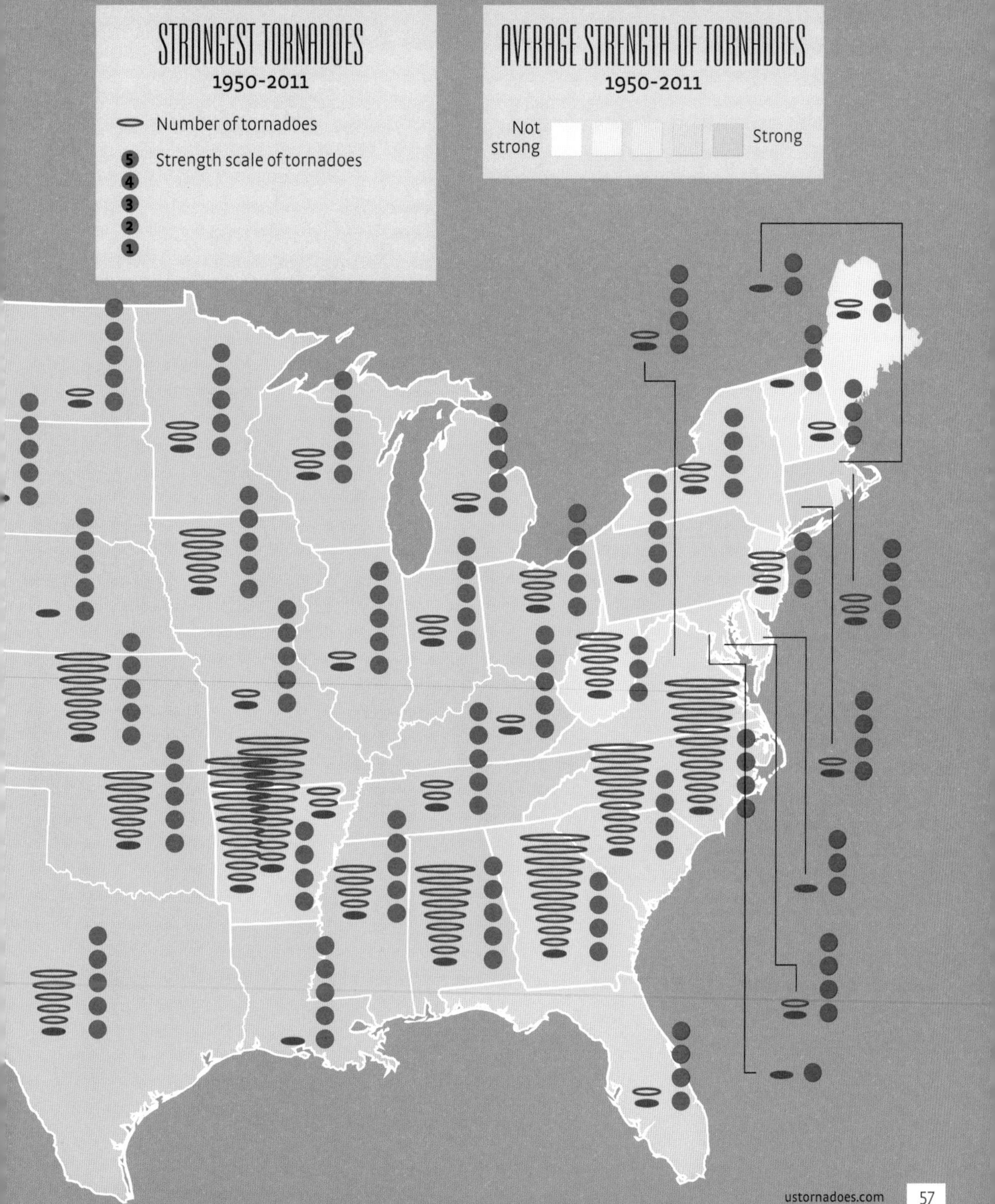
STRONGEST TORNADOES
1950-2011
Number of tornadoes
Strength scale of tornadoes
5
4
3
2
1
AVERAGE STRENGTH OF TORNADOES
1950-2011
Not strong
Strong

IT'S **A DOG'S LIFE** LINE

*The domestic dog (*Canis domesticus*) is thought to have arisen from the Grey Wolf (*Canis lupus*) between 15,000 and 30,000 years ago. The origin of all modern dog breeds is a now-extinct descendant of the Grey Wolf called* Canis lupus familiaris.

Taking just two of the oldest breeds, the Samoyed and Egyptian Greyhound, we see that they've produced breeds as diverse as the Chihuahua, Poodle, Spaniel, Beagle, Dachshund and Yorkshire Terrier – the world's tiniest dog.

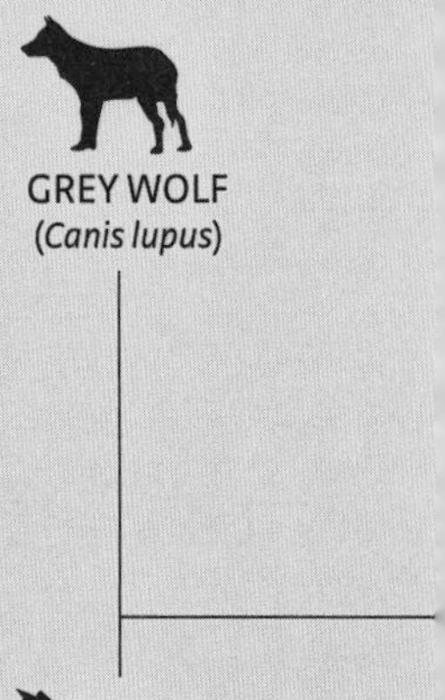

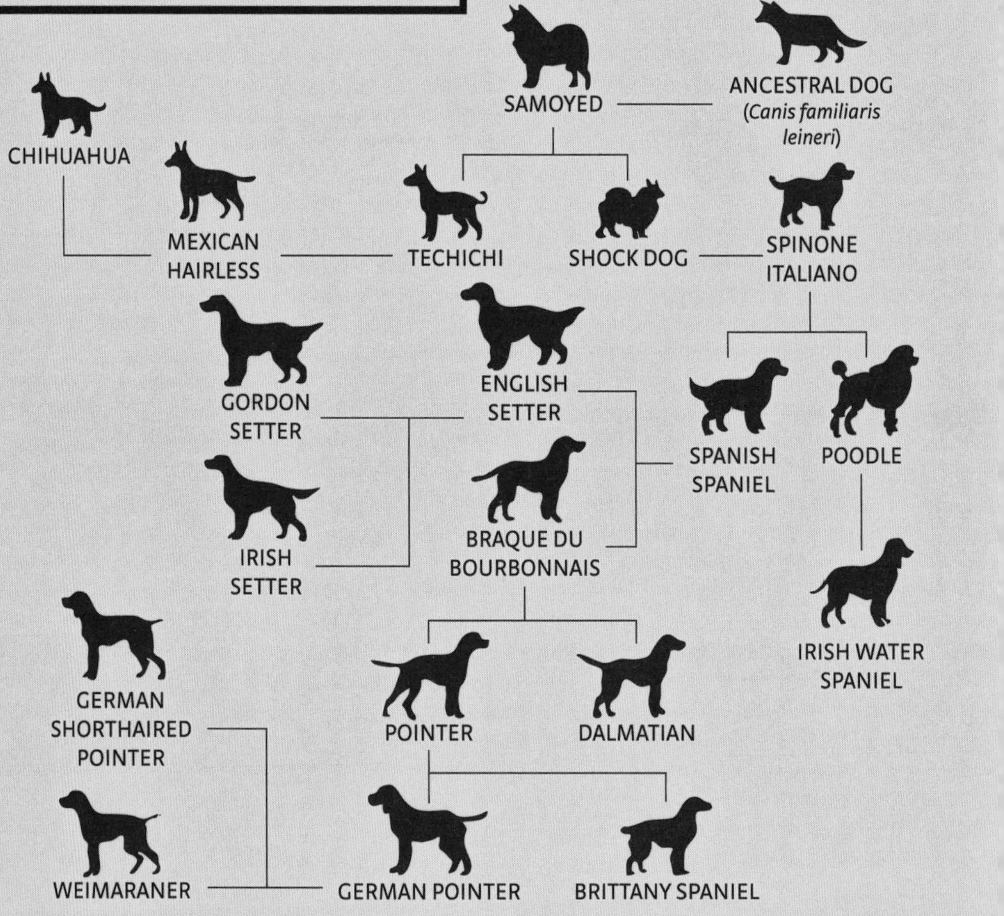

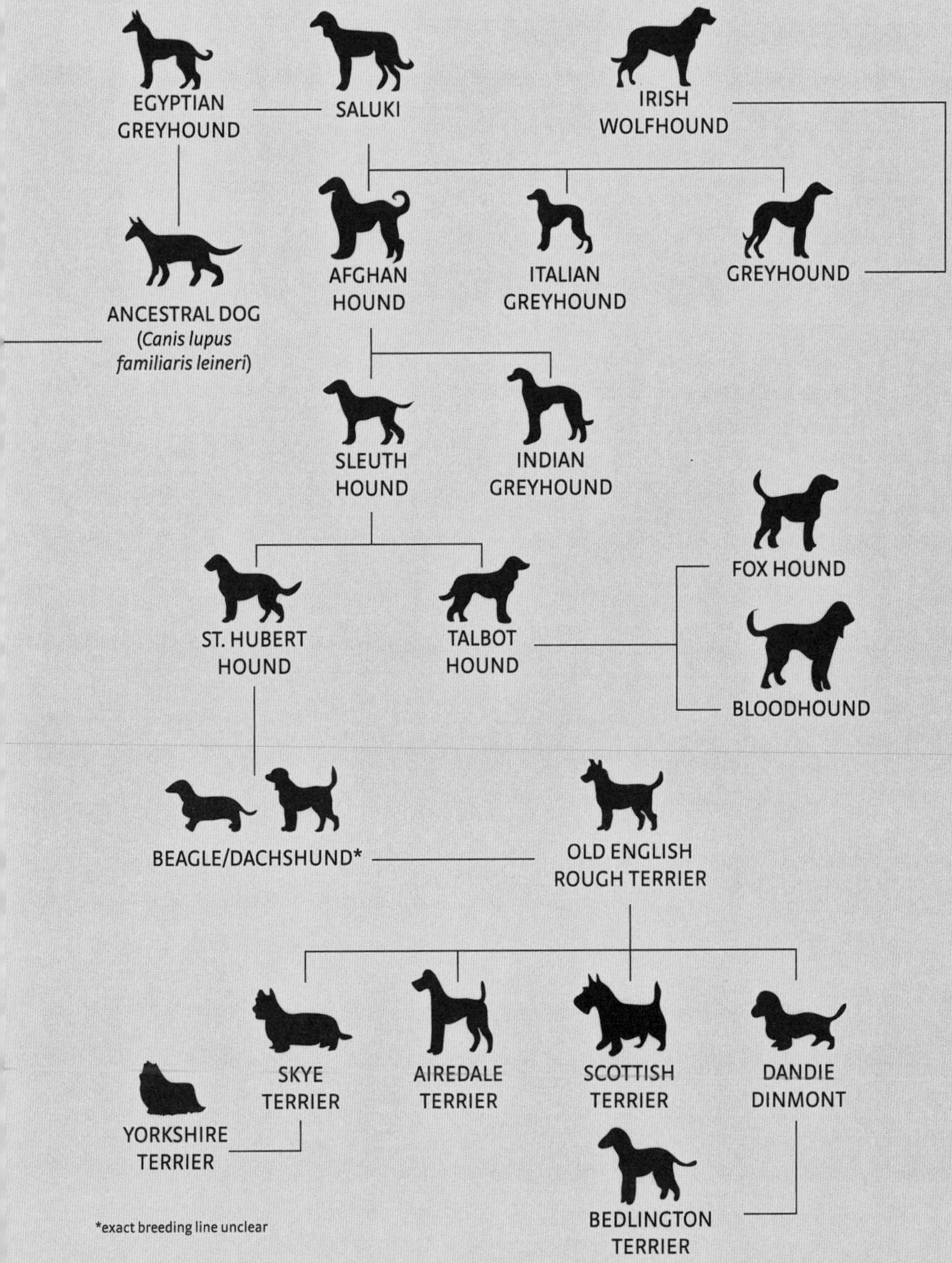
EGYPTIAN GREYHOUND
SALUKI
IRISH WOLFHOUND
ANCESTRAL DOG (Canis lupus familiaris leineri)
AFGHAN HOUND
ITALIAN GREYHOUND
GREYHOUND
SLEUTH HOUND
INDIAN GREYHOUND
FOX HOUND
ST. HUBERT HOUND
TALBOT HOUND
BLOODHOUND
BEAGLE/DACHSHUND*
OLD ENGLISH ROUGH TERRIER
SKYE TERRIER
AIREDALE TERRIER
SCOTTISH TERRIER
DANDIE DINMONT
YORKSHIRE TERRIER
BEDLINGTON TERRIER
*exact breeding line unclear

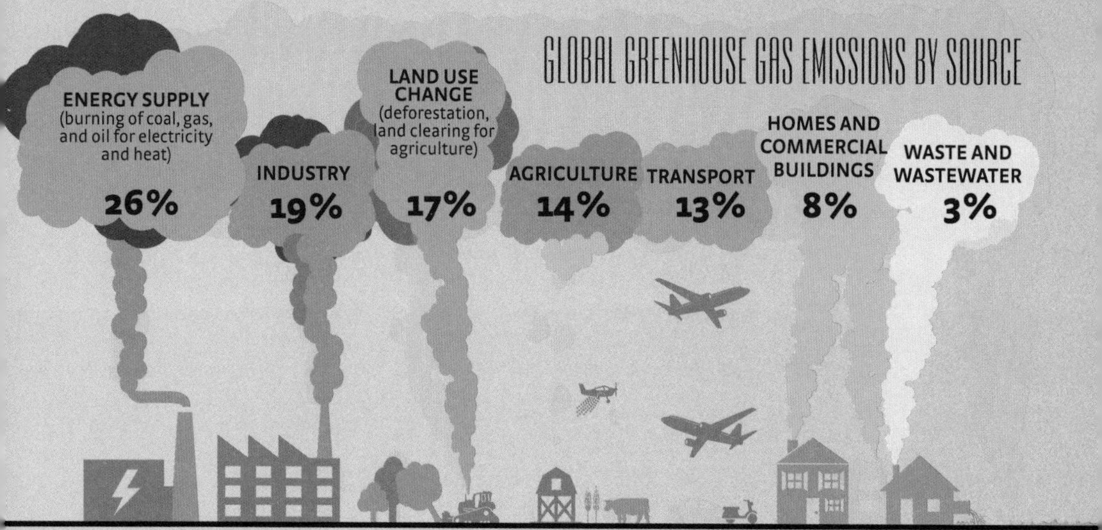

THROWING AIR BRICKS **IN A GREENHOUSE**

What makes up the 'greenhouse gases' and who's emitting the most.

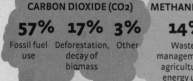

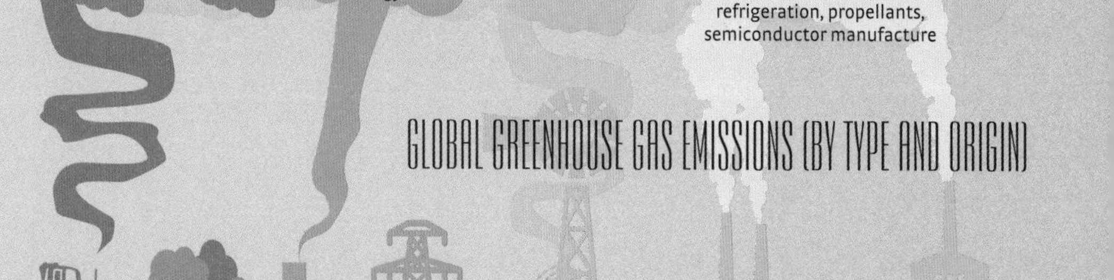

www.epa.gov, ;edgar.jrc.ec.europa.eu, Oxford Companion to Global Change

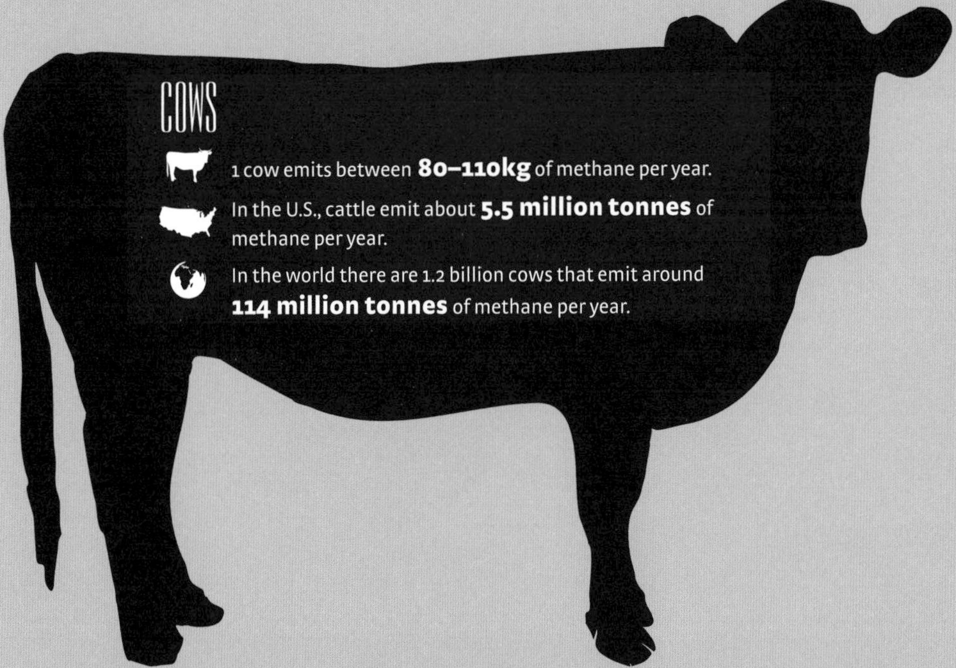

COWS

1 cow emits between **80–110kg** of methane per year.

In the U.S., cattle emit about **5.5 million tonnes** of methane per year.

In the world there are 1.2 billion cows that emit around **114 million tonnes** of methane per year.

IS IT A PLANE? IS IT A CAR? NO, **IT'S A COW**

Climate change has been blamed on commercial air travel, man's over-dependence on automobiles, and the excess number of cows bred for their meat. But which of the three contribute more harmful gases to the atmosphere over the course of a year?

BOEING 737 PASSENGER AIRLINER

(Assuming the plane does 500,000km during year)
With a 780kph cruise speed and 90kg/h CO_2 emissions, 500,000/780 = 641 hours of flight
641 × 90 = 57,690kg of CO_2 per year of travel
= **57 tonnes of CO_2 per year.**

TOYOTA PRIUS 2013 MODEL CAR

(Assuming the Prius only does 12,000km per year)
A Prius emits 49g/km CO_2 = 0.049kg/km
= **588kg of CO_2 per year.**

carbonindependent.org, epa.gov/rlep, data360.org

SHRIMPS **LOUDER** THAN A JET

As mighty as we might like to think we are, it seems that human beings don't make the loudest noise on earth. In fact, a mere snapping shrimp makes more noise just clicking its claw than one of our jet planes does at full throttle. And the acoustic bubble made by the shrimp generates temperatures as hot as the sun. Here's how man-made noises stack up in the decibel count against natural events.

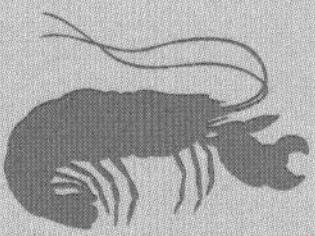

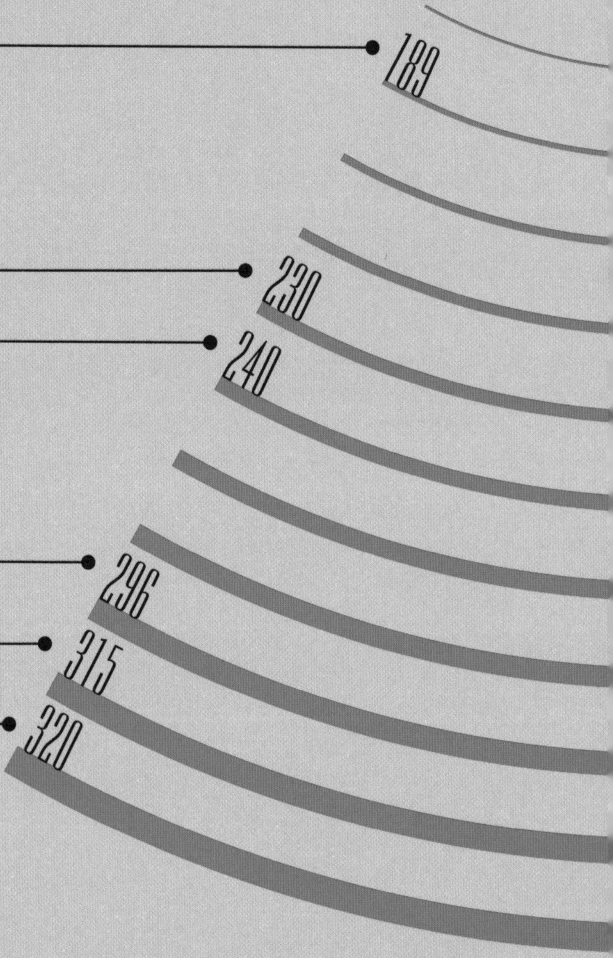

Snapping shrimp
Click from larger claw to kill fish and crabs

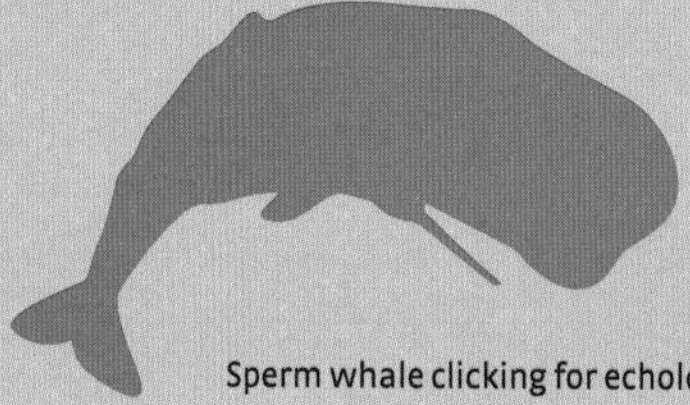

Whale
Sperm whale clicking for echolocation

Tornado
Peak of the category 5 to hit Moore, OK in 2013, at speeds of 340km/h (211mph)

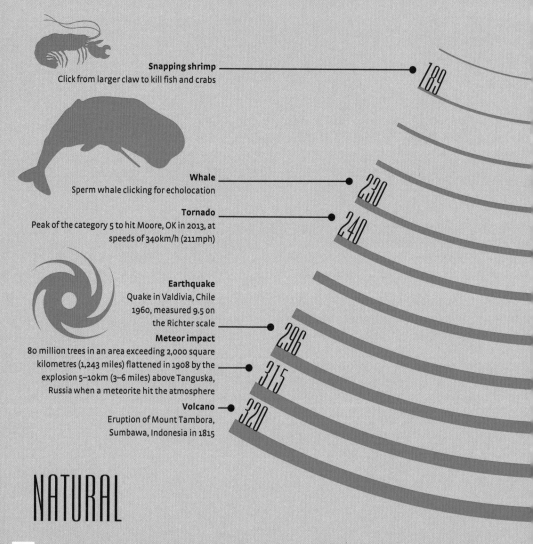

Earthquake
Quake in Valdivia, Chile 1960, measured 9.5 on the Richter scale

Meteor impact
80 million trees in an area exceeding 2,000 square kilometres (1,243 miles) flattened in 1908 by the explosion 5–10km (3–6 miles) above Tanguska, Russia when a meteorite hit the atmosphere

Volcano
Eruption of Mount Tambora, Sumbawa, Indonesia in 1815

NATURAL

DECIBEL LEVEL

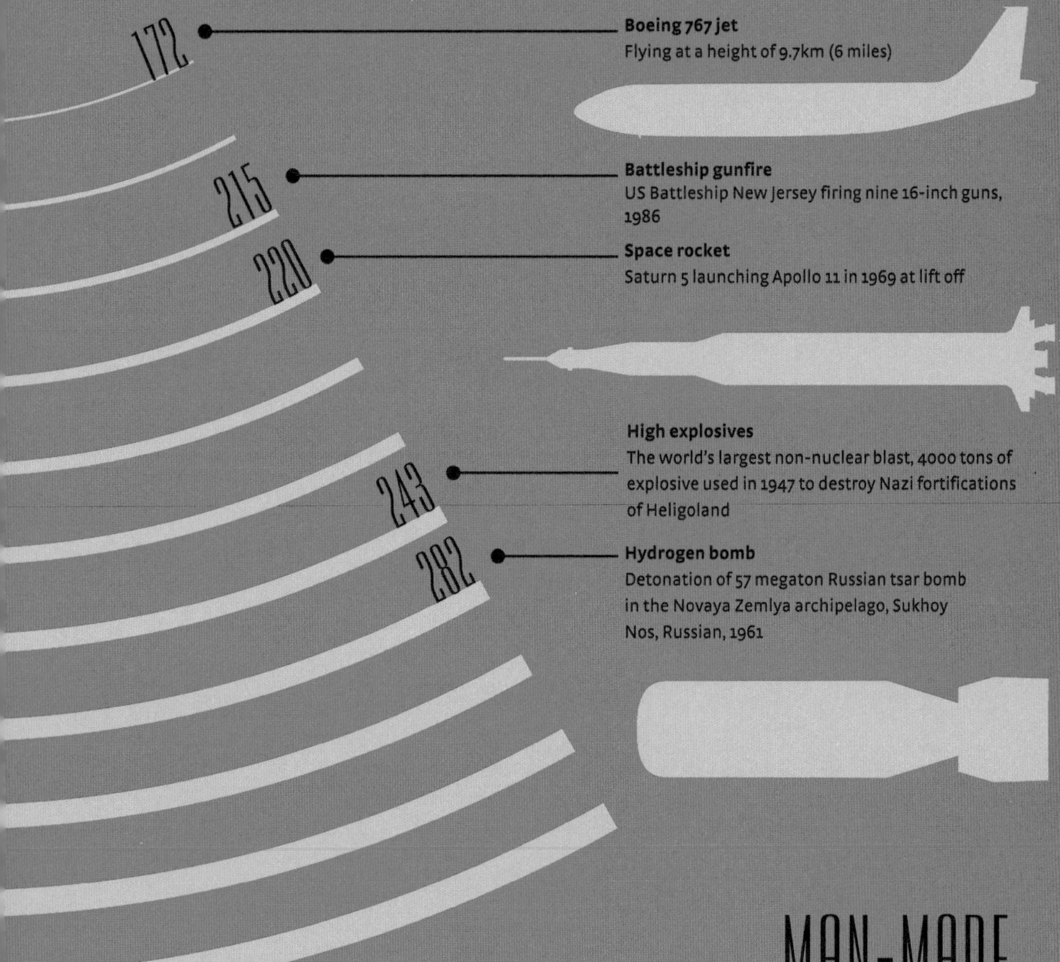

MAN-MADE

dosits.org, makeitlouder.com, wikipedia.org

EARTH TO MARS — 56 MILLION KILOMETRES / 34.8 MILLION MILES!

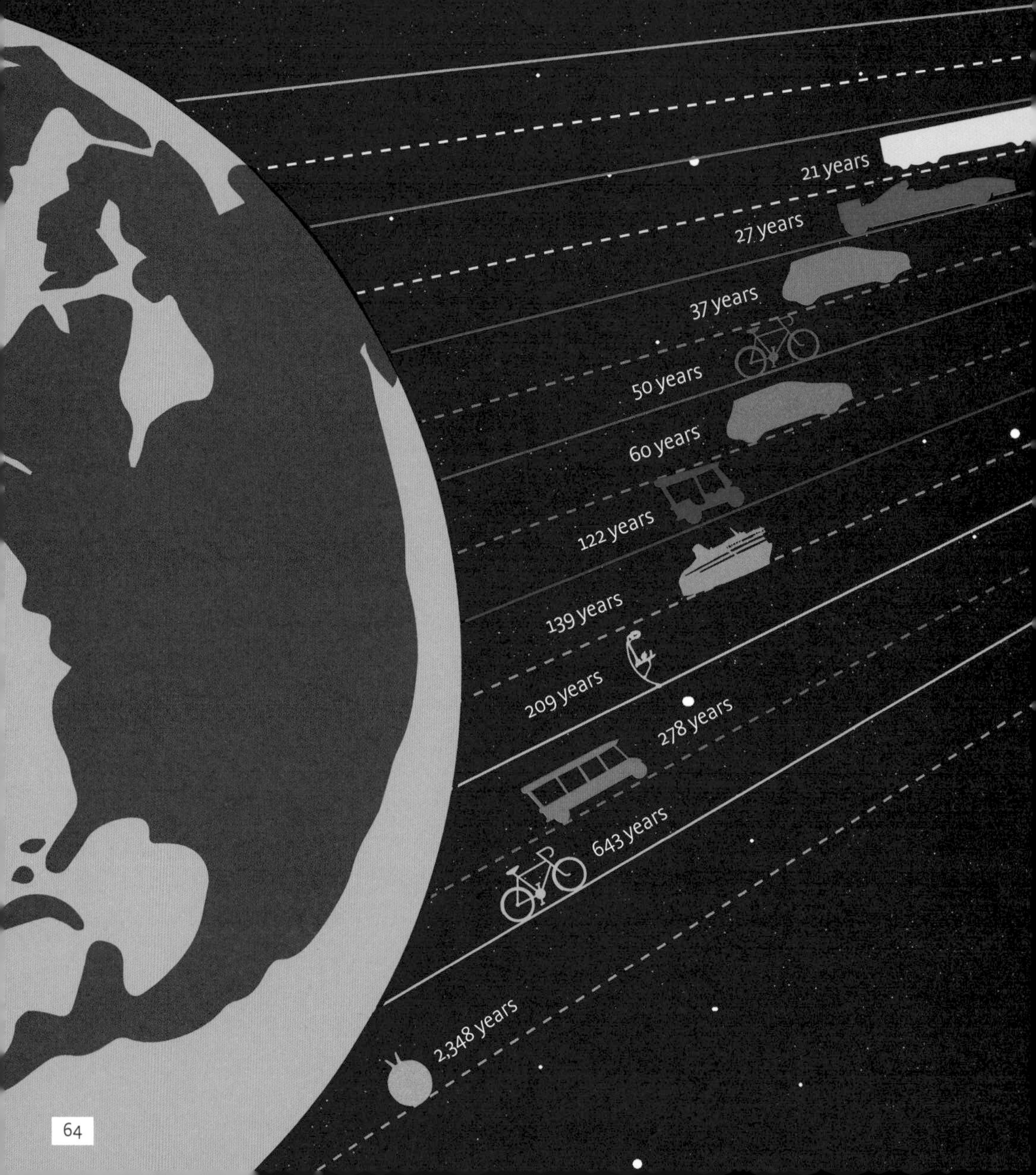

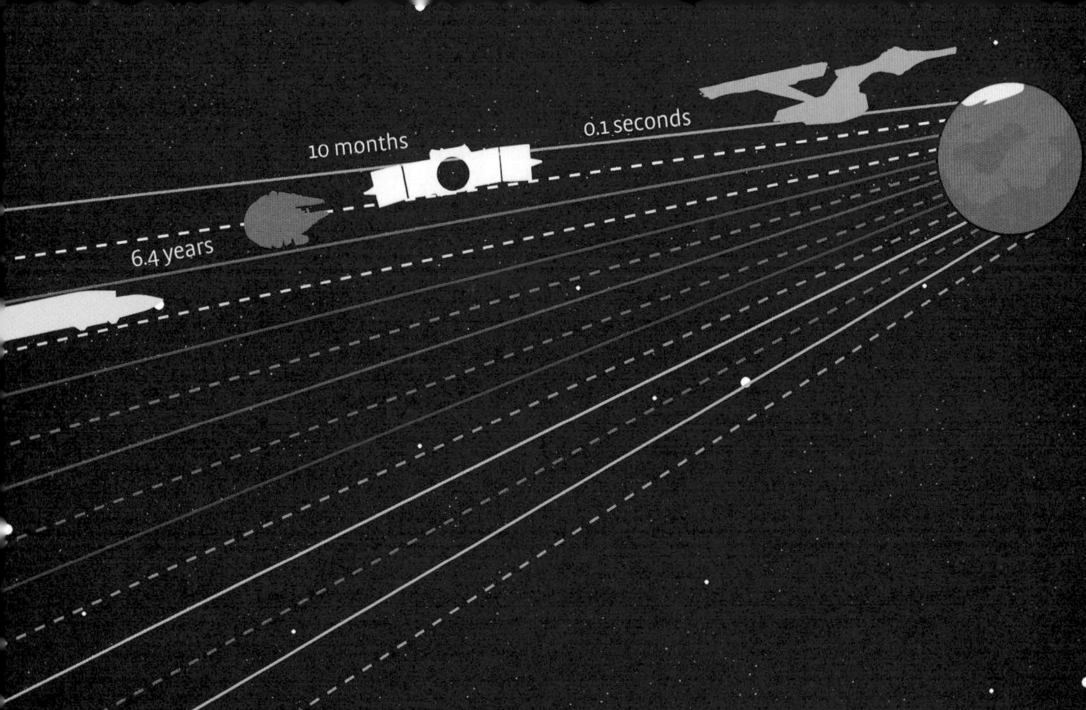

ARE WE **THERE** YET?

Now that NASA are sending probes to Mars on a regular basis, surely it can't be long before anyone can get there. Here, for the ambitious traveller, is an estimation of how long it might take to reach the red planet using different methods of transport.

- *U.S.S Enterprise*
- NASA's MAVEN
- *Millennium Falcon*
- Japanese Bullet Train
- Formula One (Ferrari)
- Toyota Prius (top speed)
- Bicycle (Streamline)
- Toyota Prius (highway limit)
- Tuk Tuk
- Cruise Ship (***Queen Mary II*** at normal cruising speed)
- Powerisers
- Pedal-powered pub cycle
- Bicycle (commuting, average)
- Space Hopper

pace.com, muse.jpl.nasa.gov, universetoday.com, toyota.pt, japan-guide.com, cunard.com, startrek.com, bikeforums.net, rickshawchallenge.com, poweriser.tribe.net, wikipedia.org

BURNING UP IN YOUR KITCHEN

The various innocuous-looking electrical items scattered about the average kitchen are all capable of burning up units of electricity, and some more efficiently than others. If you want to save money on your energy bills, wash clothes by hand.

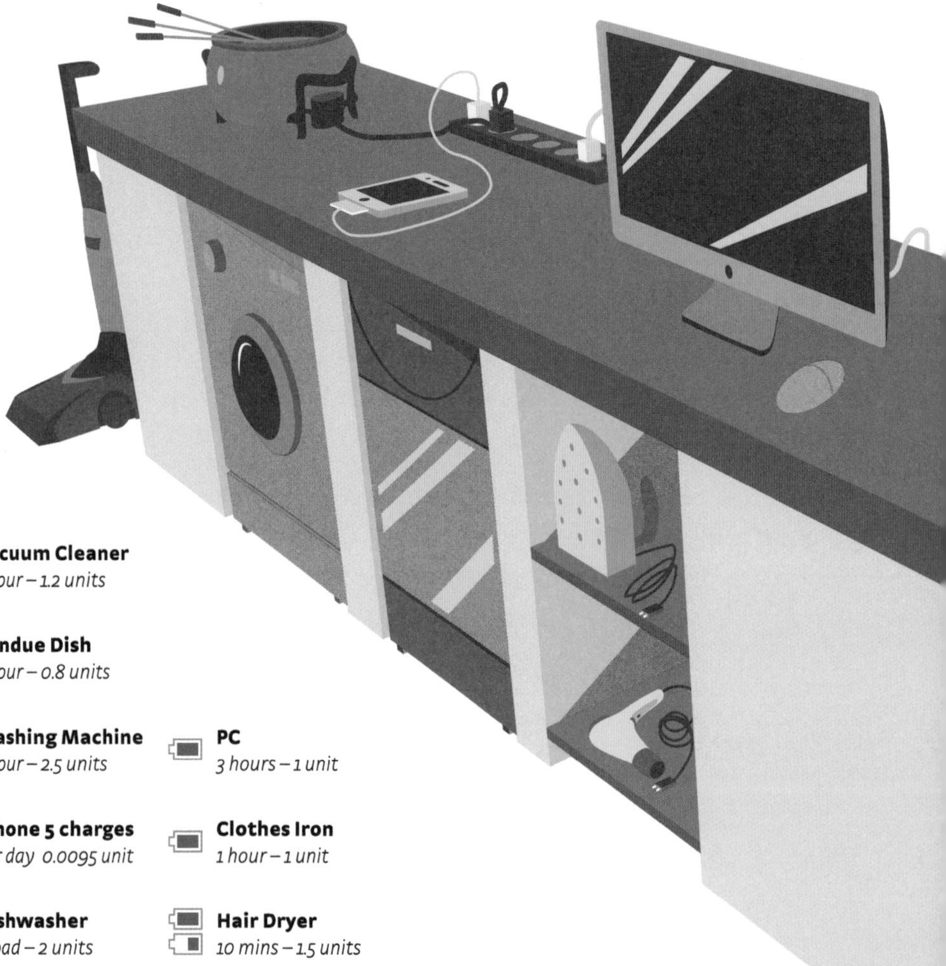

Vacuum Cleaner
1 hour – 1.2 units

Fondue Dish
1 hour – 0.8 units

Washing Machine
1 hour – 2.5 units

iPhone 5 charges
Per day 0.0095 unit

Dishwasher
1 load – 2 units

PC
3 hours – 1 unit

Clothes Iron
1 hour – 1 unit

Hair Dryer
10 mins – 1.5 units

(1 unit = 1 kilowatt-hour (kWh) or 1000 watts of power used for one hour)

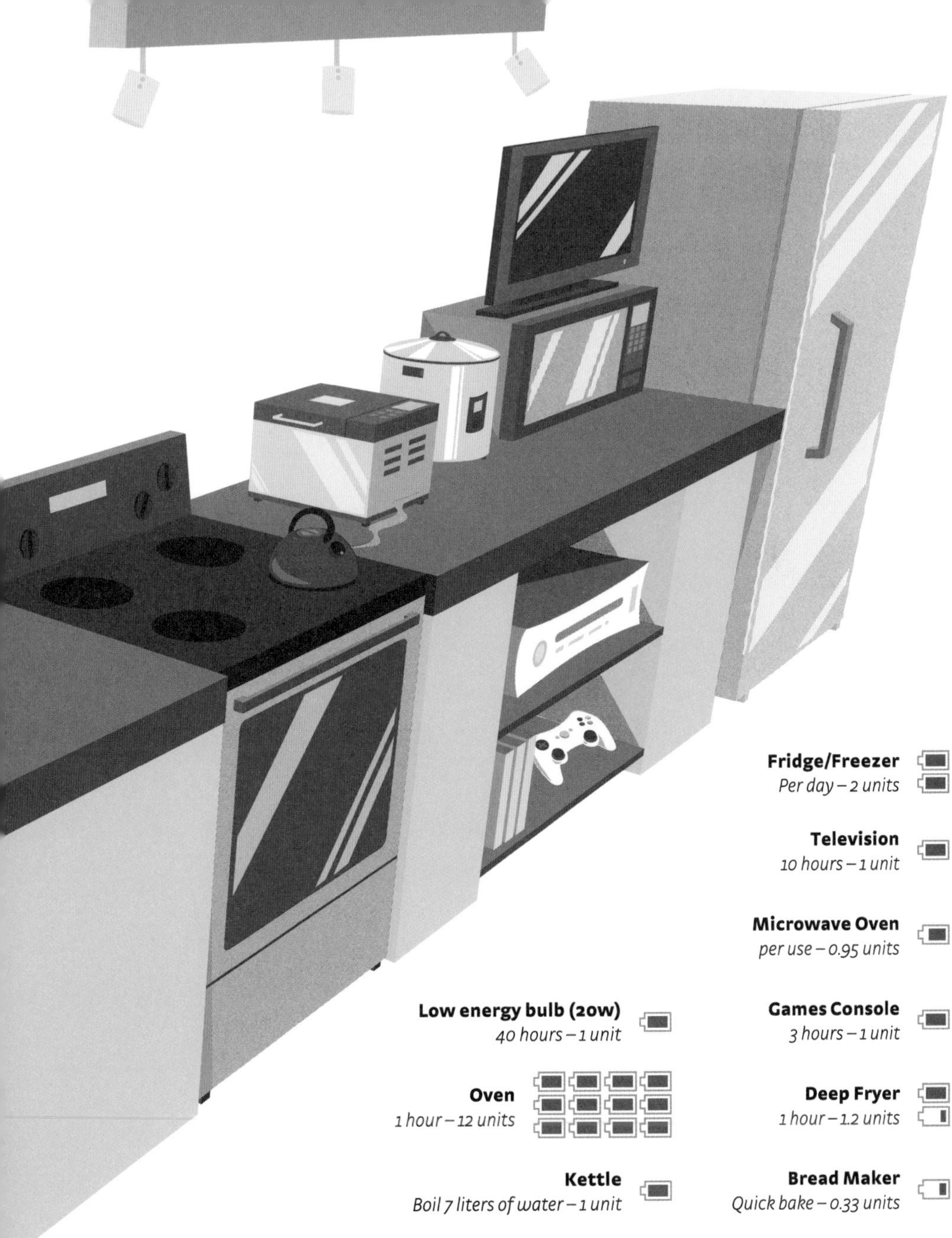

Fridge/Freezer
Per day – 2 units

Television
10 hours – 1 unit

Microwave Oven
per use – 0.95 units

Games Console
3 hours – 1 unit

Deep Fryer
1 hour – 1.2 units

Bread Maker
Quick bake – 0.33 units

Low energy bulb (20w)
40 hours – 1 unit

Oven
1 hour – 12 units

Kettle
Boil 7 liters of water – 1 unit

Deadliest earthquakes of **all-time** and of the **21st century**

Of all-time | Of the 21st century

Location (date)	Year	# deaths	magnitude richter scale (0–10)
1 Shaanxi, China	**1556**	830,000	
2 Haiti	**2010**	316,000	
3 Tangshan, China	**1976**	242,000	
4 Aleppo, Syria	**1138**	230,000	n/a
5 Sumatra, Indonesia	**2004**	228,000	
6 Damghan, Iran	**856**	200,000 (est)	n/a
7 Haiyuan, China	**1920**	200,000	
8 Ardabil, Iran	**893**	150,000 (est)	n/a
9 Kanto, Japan	**1923**	143,000 (est)	
10 Ashgabat, Turkmenistan	**1948**	110,000	

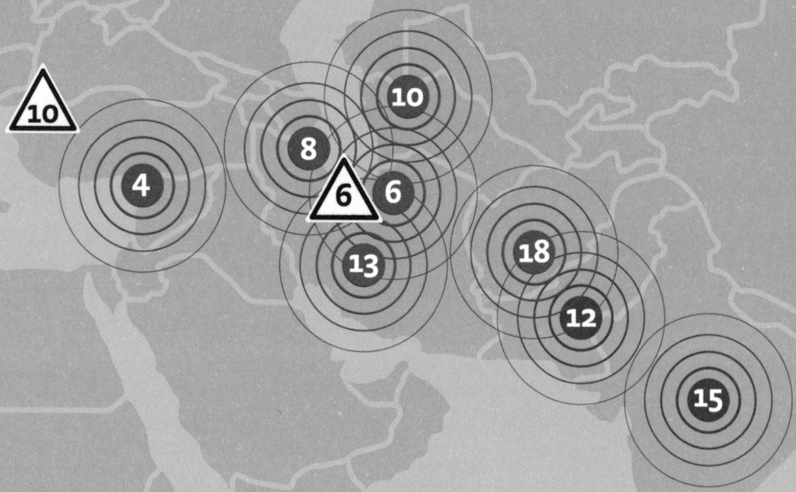

Location (date)	Year	# deaths	magnitude richter scale (0–10)
11 Szichuan, China	**2008**	87,600	
12 Pakistan	**2005**	80,400	
13 Iran	**2003**	31,000	
14 Honshu, Japan	**2011**	21,000	
15 India	**2001**	20,000	
16 Java, Indonesia	**2006**	20,000	
17 Sumatra, Indonesia	**2009**	1,117	
18 Afghanistan	**2002**	1,000	

Largest populations at risk of earthquake damage

	City	country	area	population
1	Tokyo-Yokohama	Japan	16,300 km^2	29.4 million
2	Jakarta	Indonesia	11,600 km^2	17.7 million
3	Manila	Philippines	2,900 km^2	16.8 million
4	Los Angeles	USA	14,400 km^2	14.7 million
5	Osaka-Kobe	Japan	13,600 km^2	14.6 million
6	Tehran	Iran	11,000 km^2	13.6 million
7	Nagoya	Japan	15,600 km^2	9.4 million
8	Lima	Peru	2,600 km^2	8.9 million
9	Taipei	Taiwan	2,100 km^2	8.0 million
10	Istanbul	Turkey	4,100 km^2	6.4 million

QUAKING IN OUR BEDS

The most volatile seismic regions of our planet lie on fault lines, where the Earth's continental plates meet. Unfortunately, these are also the locations of some of our greatest population masses. Here are the metropolitan areas most under threat, plus the deadliest quakes of this century and in history.

earthquake.usgs.gov, livescience.com, viewsoftheworld.com, fastcoexist.com, media.swissre.com

PANDEMIC **VS** WAR

Mother Nature has always found ways of cutting back on the human population, and from the Middle Ages to present day a variety of pandemics have swept through the species. Man has perhaps done a better job of cutting back on the (mostly male) population in wars.

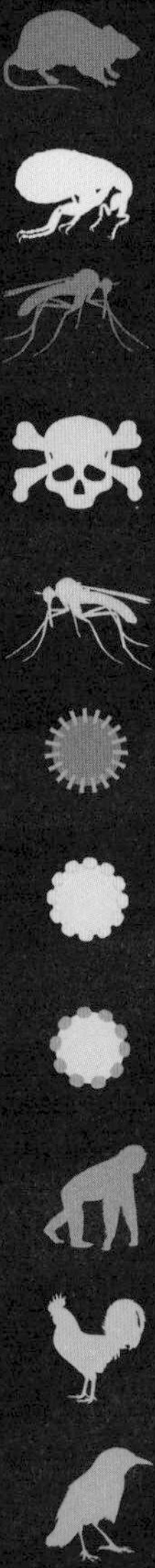

Bubonic Plague of Justinian
(541–750 AD)
25–100 million

The Black Plague
(1347–1351) 25 million

Malaria
(1600–present)
826 million

Bubonic Plague
London (1665–1666)
75,000–100,000

Yellow Fever
Philadelphia (1793)
5,000 deaths

Smallpox
(1900–1980)
300–500 million

Typhus Eastern Europe
(1914–1915)
million+

Influenza
(1918–1919)
40 million

AIDS
(1981–present)
25 million +

SARS
China (2009)
755

Avian Influenza
(H1N1)
380

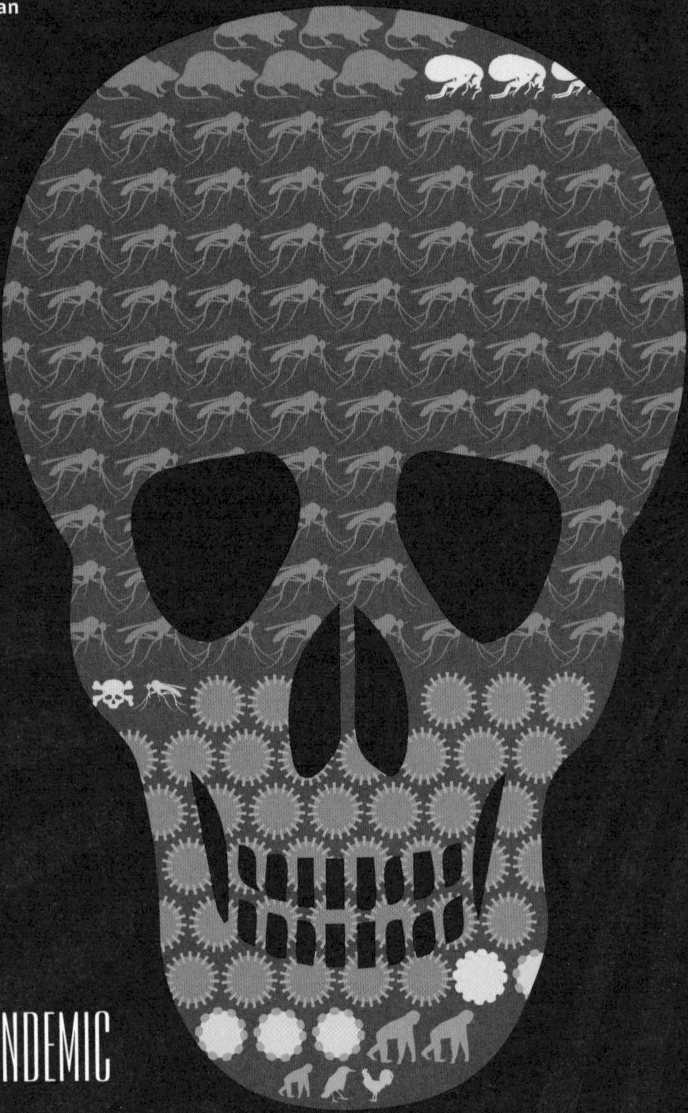

PANDEMIC

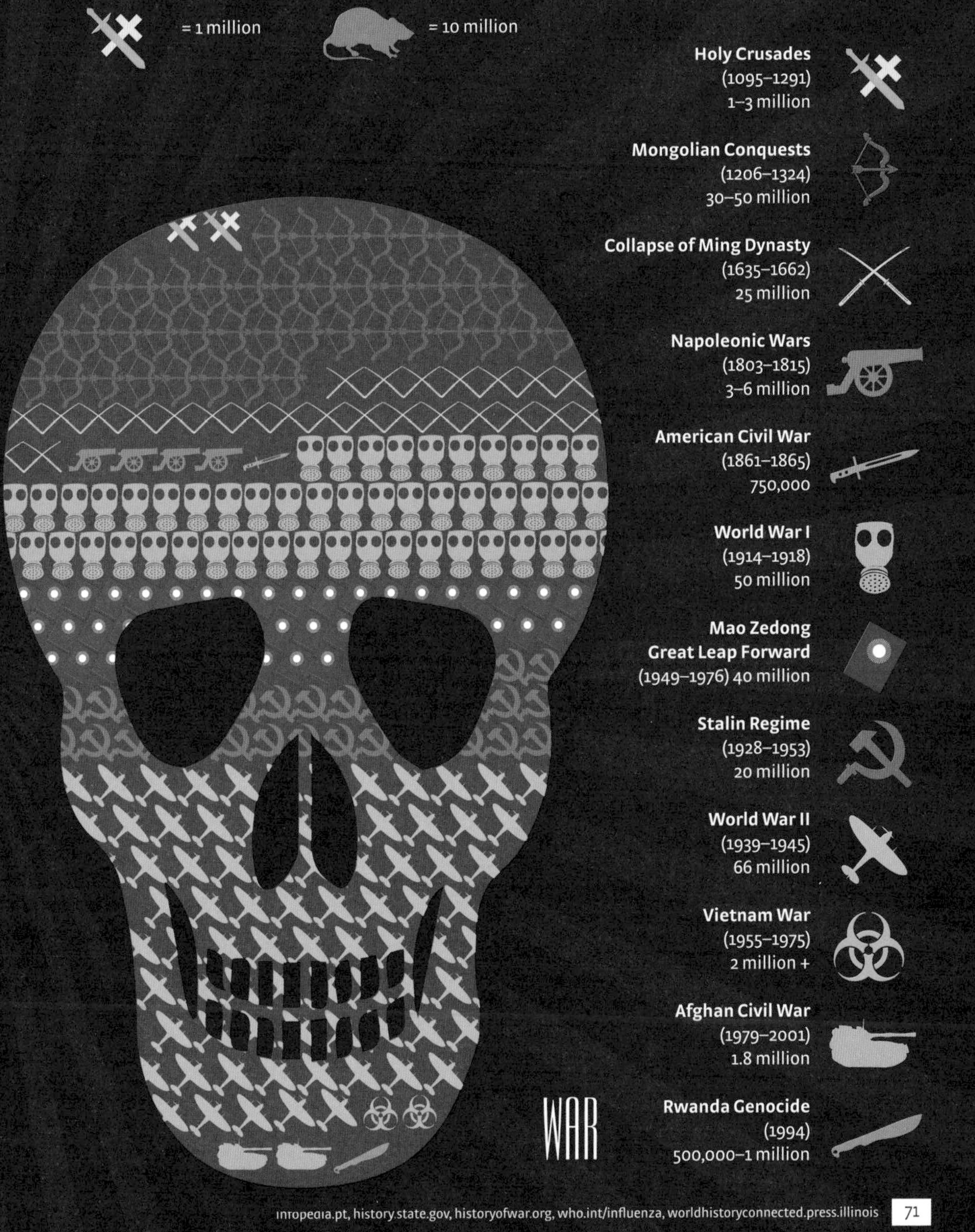

infopedia.pt, history.state.gov, historyofwar.org, who.int/influenza, worldhistoryconnected.press.illinois

SOME BRAINS ARE
BIGGER THAN OTHERS

Pound for pound an adult human brain is a thousand times the size that of a hamster, but its ratio of weight to body weight is far smaller. So you could say that a hamster has a bigger brain than we do.

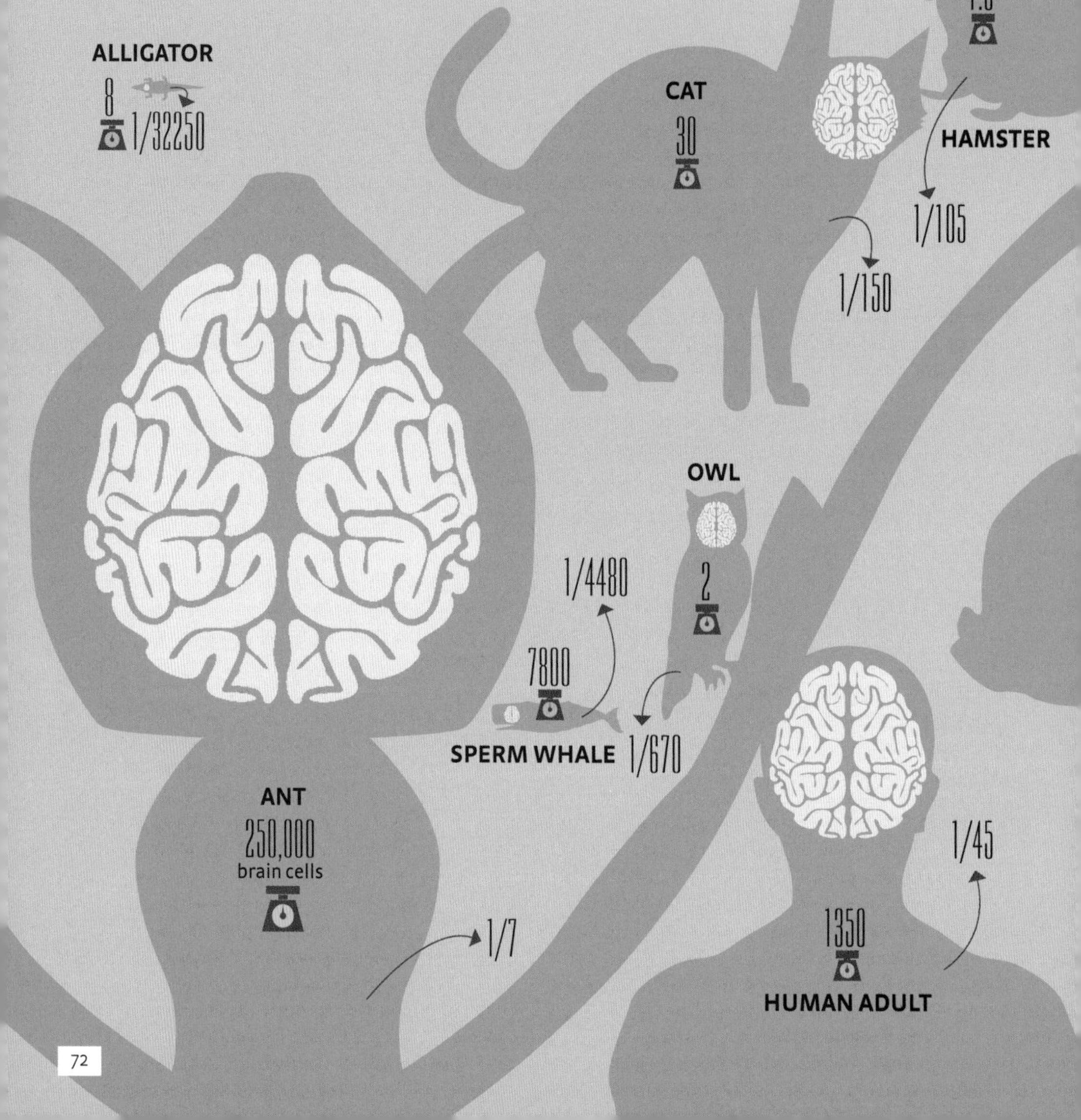

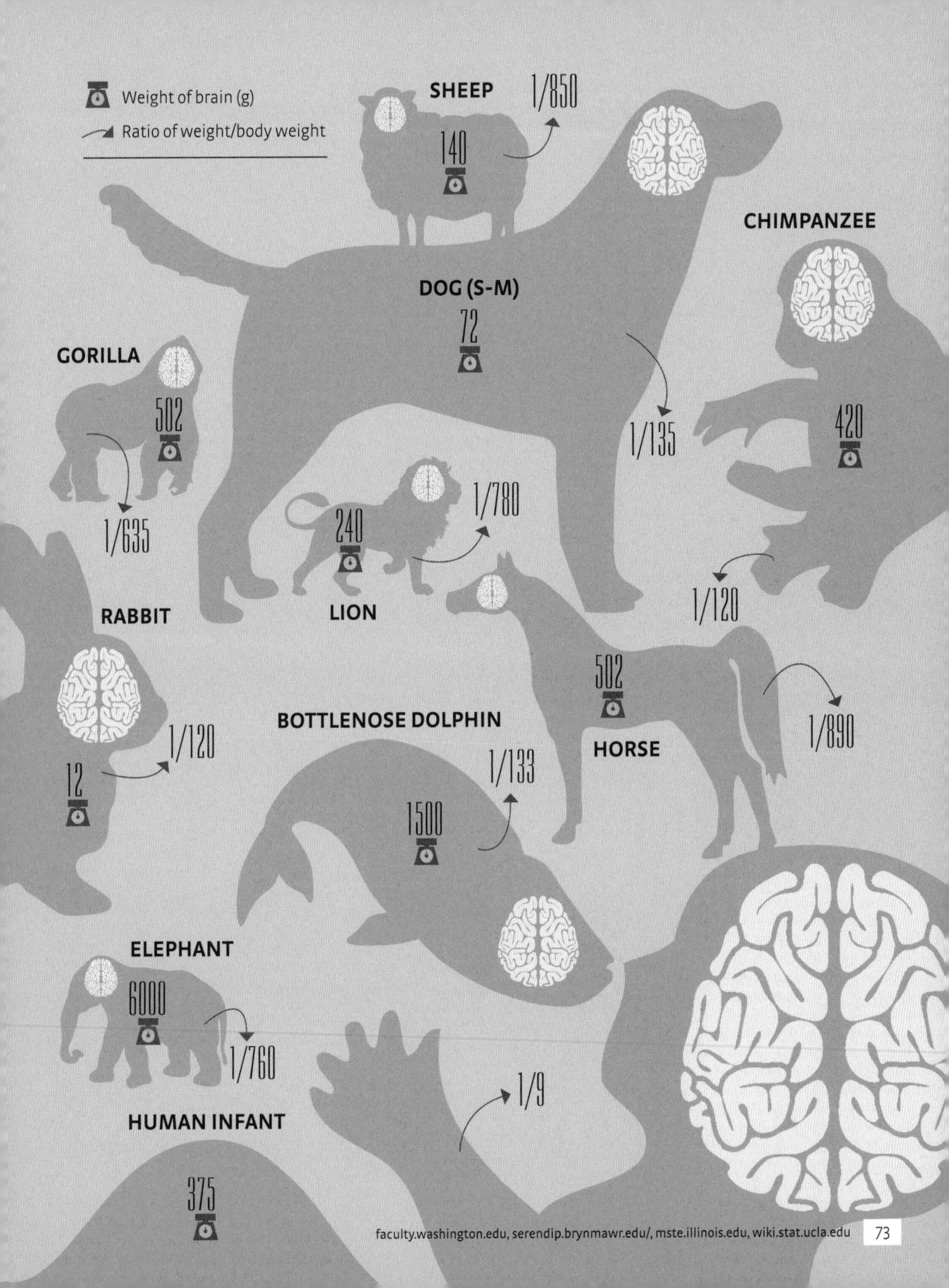

faculty.washington.edu, serendip.brynmawr.edu/, mste.illinois.edu, wiki.stat.ucla.edu

THE **CARBON FOOTPRINT** DIET

It's no wonder that many vegetarians are so ecologically minded, their diet has a far less harmful impact on the environment than that of meat eaters. Here's the carbon footprint for five different diets.

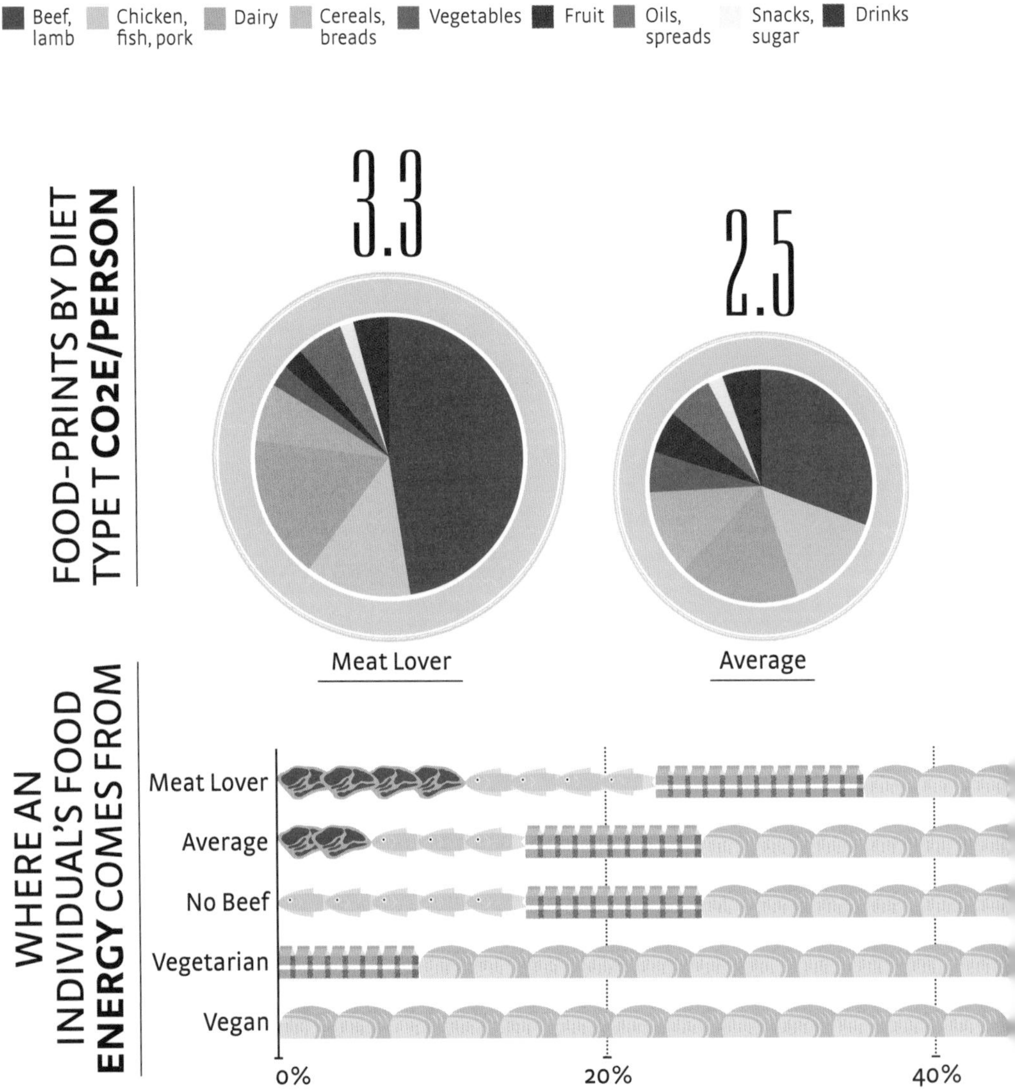

CARBON INTENSITY OF EATING: **G CO2/KCAL**

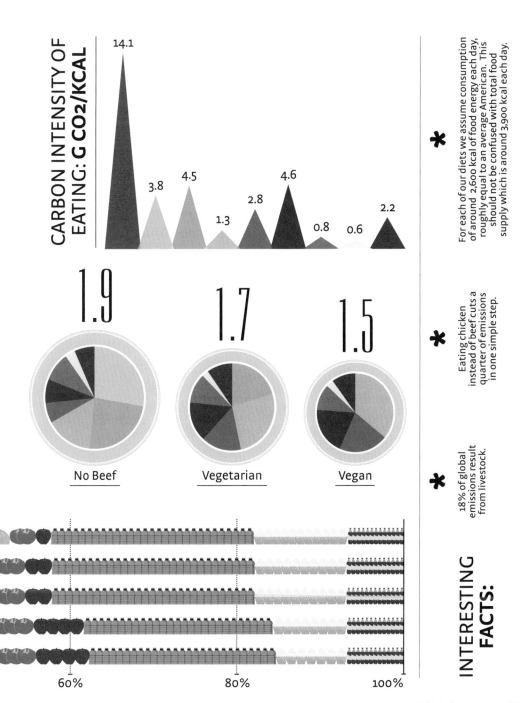

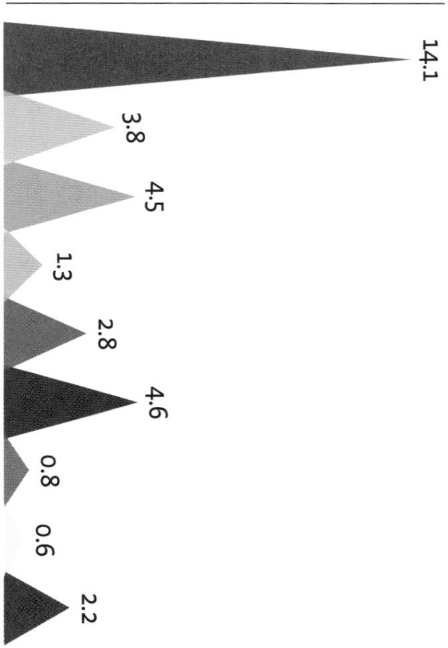

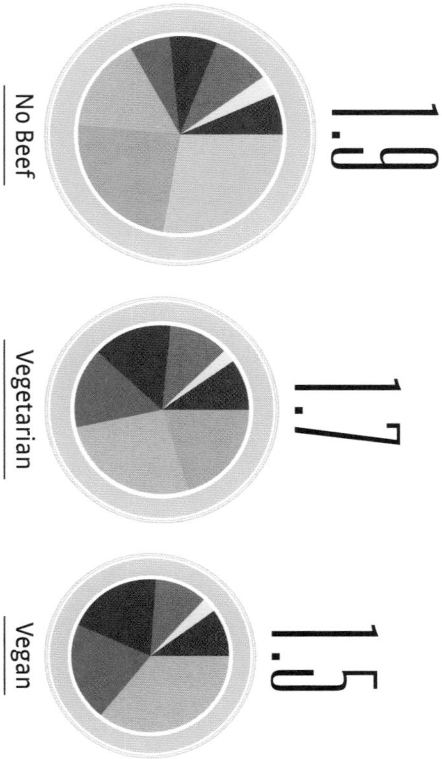

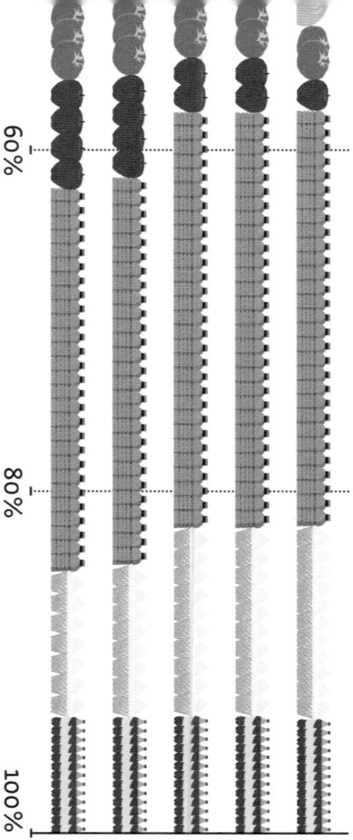

INTERESTING **FACTS:**

* 18% of global emissions result from livestock.

* Eating chicken instead of beef cuts a quarter of emissions in one simple step.

* For each of our diets we assume consumption of around 2,600 kcal of food energy each day, roughly equal to an average American. This should not be confused with total food supply which is around 3,900 kcal each day.

VACUUM PACKED **WORLD**

In one of the largest surveys of its kind, in 2013 vacuum cleaner manufacturer Electrolux asked 28,000 people in 23 different countries about their vacuuming habits. The results were surprising in many ways, not least concerning which nations preferred to vacuum while naked, what age groups were regular cleaners and when vacuuming was done.

MEN vs WOMEN

Who vacuums more in the battle of the sexes?

The answers could be closer than you think!

Vacuum once a day?	Vacuum 2-5 times a week?	Vacuum once a month?

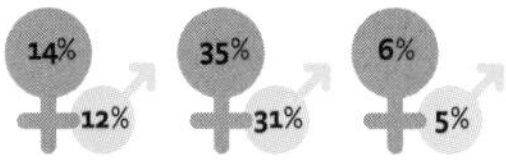

Which country keeps the **CLEANEST?**

Vacuum once a day

CHILE 31%

Global avg 13%

Vacuum once a month

CHINA 17%

Global avg 6%

However, **Brazilians** vacuum the longest

Brazil avg.

1-2 HOURS

in a vacuuming session

NAKED

NORWAY

Surprisingly

Vacuum more in the nude than any other country

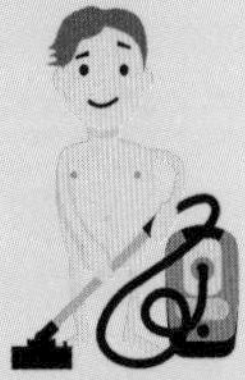

BUT

Unsurprisingly

More men vacuum in the nude than women

What people **WEAR**

Casual	**Tracksuit**	**Workwear**	**Underwear**	**Nude**
69%	**21%**	**5%**	**4%**	**2%**

HOUSE CLEANING HOBBIES

What people **DO**

51% of people turn up their favourite tune

But only 3% of people admit they dance away the dust

27% of people think of nothing at all

Whilst only **3%** pass the time with problem solving

PILLS FOR ALL ILLS

Advances in the development of medicine and increasing longevity mean that the world-wide market for prescription drugs continues to grow. What are the biggest markets and the most popular pills in the multi-billion dollar pharma market?

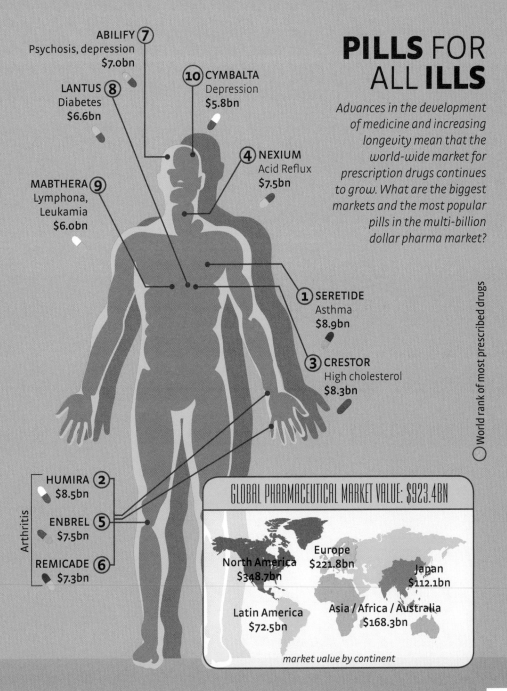

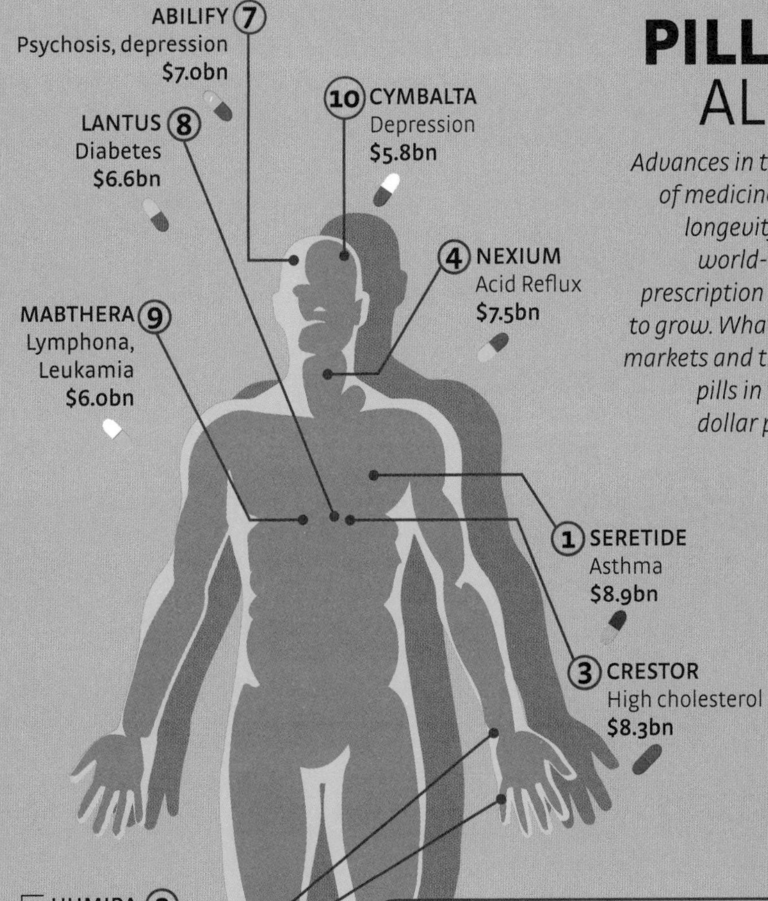

MAKE WORDS **OR WARS?**

Comparing the top 15 countries by annual expenditure on the military (by amount) with the figures that each spends on education we get a good idea of what each nation's leaders see as more important. When placing the highest military spenders by the amount they spend in the world rankings for education, none would make the top 30.

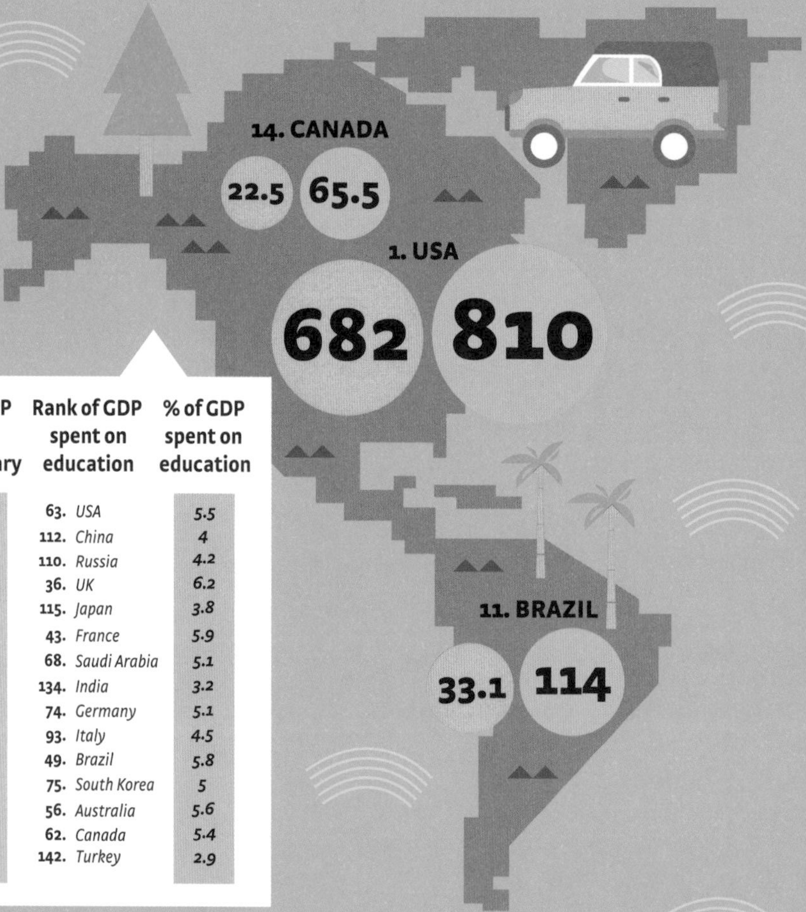

Country by GDP Overall	% of GDP spent on military	Rank of GDP spent on education	% of GDP spent on education
1. *USA*	4.4	63. *USA*	5.5
2. *China*	2	112. *China*	4
3. *Russia*	4.4	110. *Russia*	4.2
4. *UK*	2.5	36. *UK*	6.2
5. *Japan*	1	115. *Japan*	3.8
6. *France*	2.3	43. *France*	5.9
7. *Saudi Arabia*	8.9	68. *Saudi Arabia*	5.1
8. *India*	2.5	134. *India*	3.2
9. *Germany*	1.4	74. *Germany*	5.1
10. *Italy*	1.7	93. *Italy*	4.5
11. *Brazil*	1.5	49. *Brazil*	5.8
12. *South Korea*	2.7	75. *South Korea*	5
13. *Australia*	1.7	56. *Australia*	5.6
14. *Canada*	1.3	62. *Canada*	5.4
15. *Turkey*	2.3	142. *Turkey*	2.9

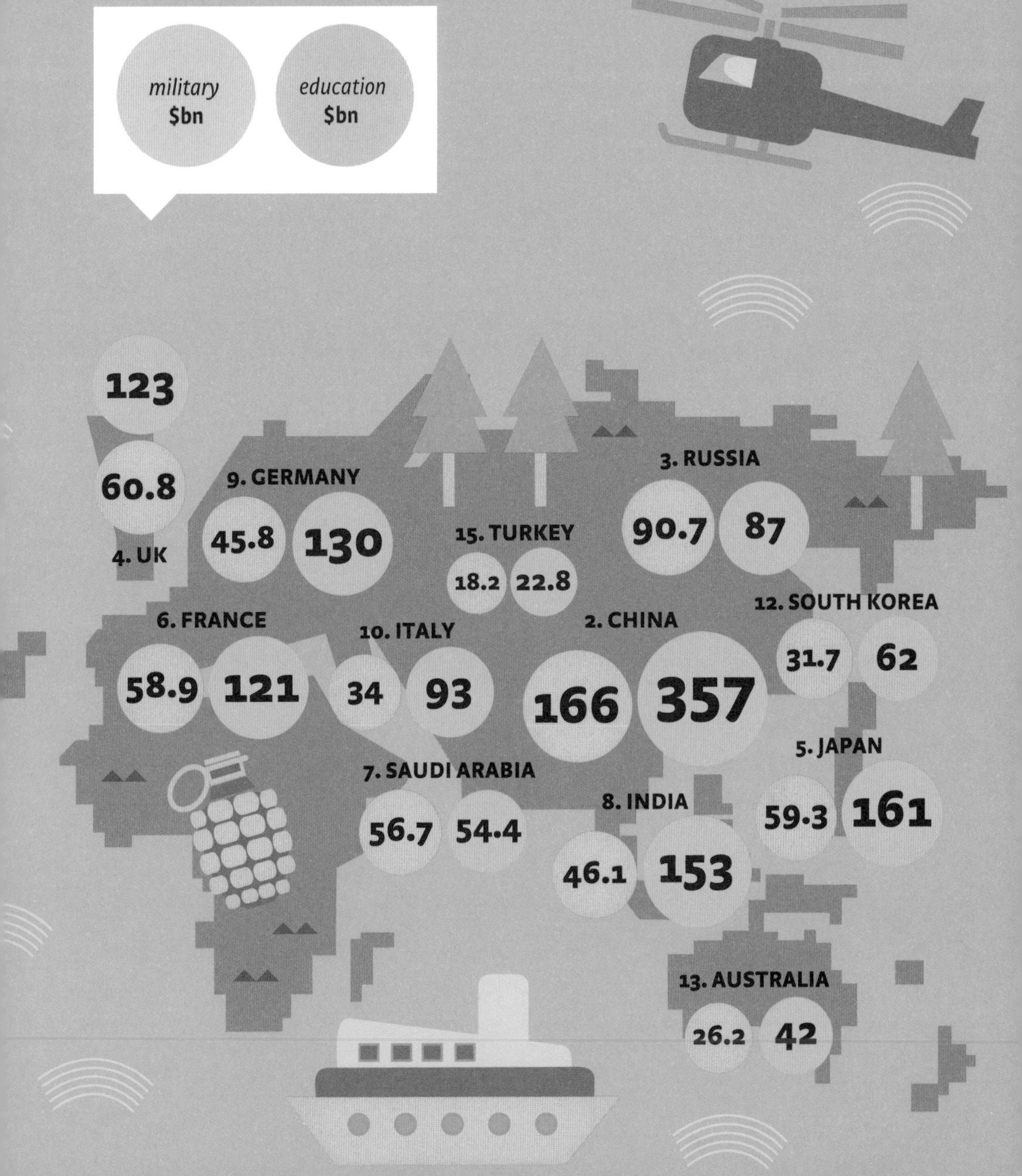

wikidepia.org, cia.gov, countryeconomy.com

EXTREME **TASTE TEST**

TASTE		NAME	FOUND IN
BITTER		Quassin	quassia (bitter wood) tree
		Denatonium Benzoate (Bitrex)	supermarkets everywhere
SMELLY		Butyl seleno-mercaptan	skunk spray
		Ethyl mercaptan	fuel stores
SWEET		Thaumatin	katemfe fruit of West Africa
		Lugduname	laboratory only
SOUR		Fumaric acid	fumitiory plant
		Malic acid	fast food restaurants
POISONOUS		Botulinum toxin A	bacterium clostridium botulinum
		Sarin	military compounds

Chemists and scientists have tried to best nature in creating artificial extreme tastes, smells, poisons and flavouring for centuries. But as this graphic shows, nature always wins. Comparing the smelliest, most sour, sweetest, most bitter and most poisonous matter found in nature and from the laboratory.

Natural

Man-made

DETAIL	USE
detectable at 0.08 parts per million	herbal medicines
tasted at 0.05 parts per million	in anti-freeze, wood alcohol, soap, household cleaners etc
there is nothing stronger	warding off attack
causes nausea, headaches and possible organ failure	added to propane gas to indicate gas leaks
3,000 times sweeter than table sugar (sucrose)	food stuff additive, including chewing gum
220,000 times sweeter than sucrose	not licensed for use
twice as sour as citric acid	food flavouring
deadliest natural toxin, toxicity level of LD50 (will kill 50% test subjects) at 0.00000003 mg per kg	chemical weapons
LD50 of 0.4mg per kg	

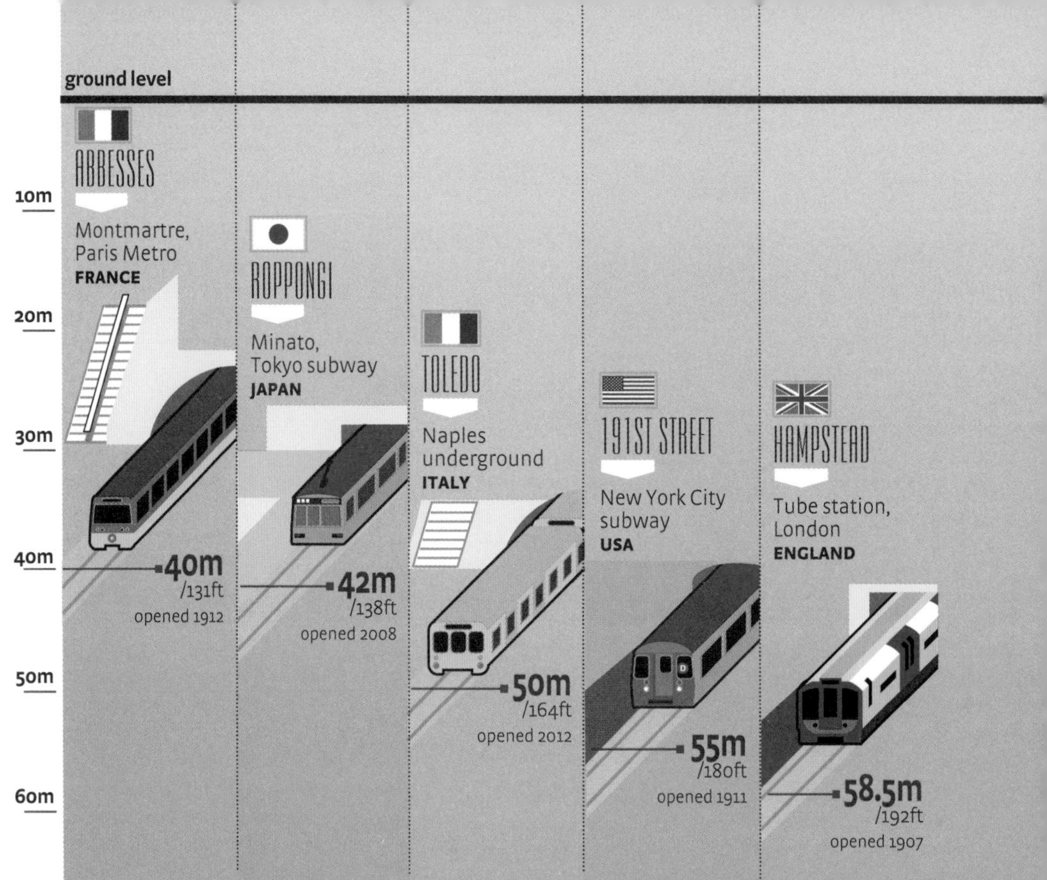

HOW **LOW** CAN YOU GO?

Travelling underground in any of the world's great cities can feel like descending into the depths of the earth. In these cities, and at these stations, catching a train feels even more so. These are the world's deepest subway stations.

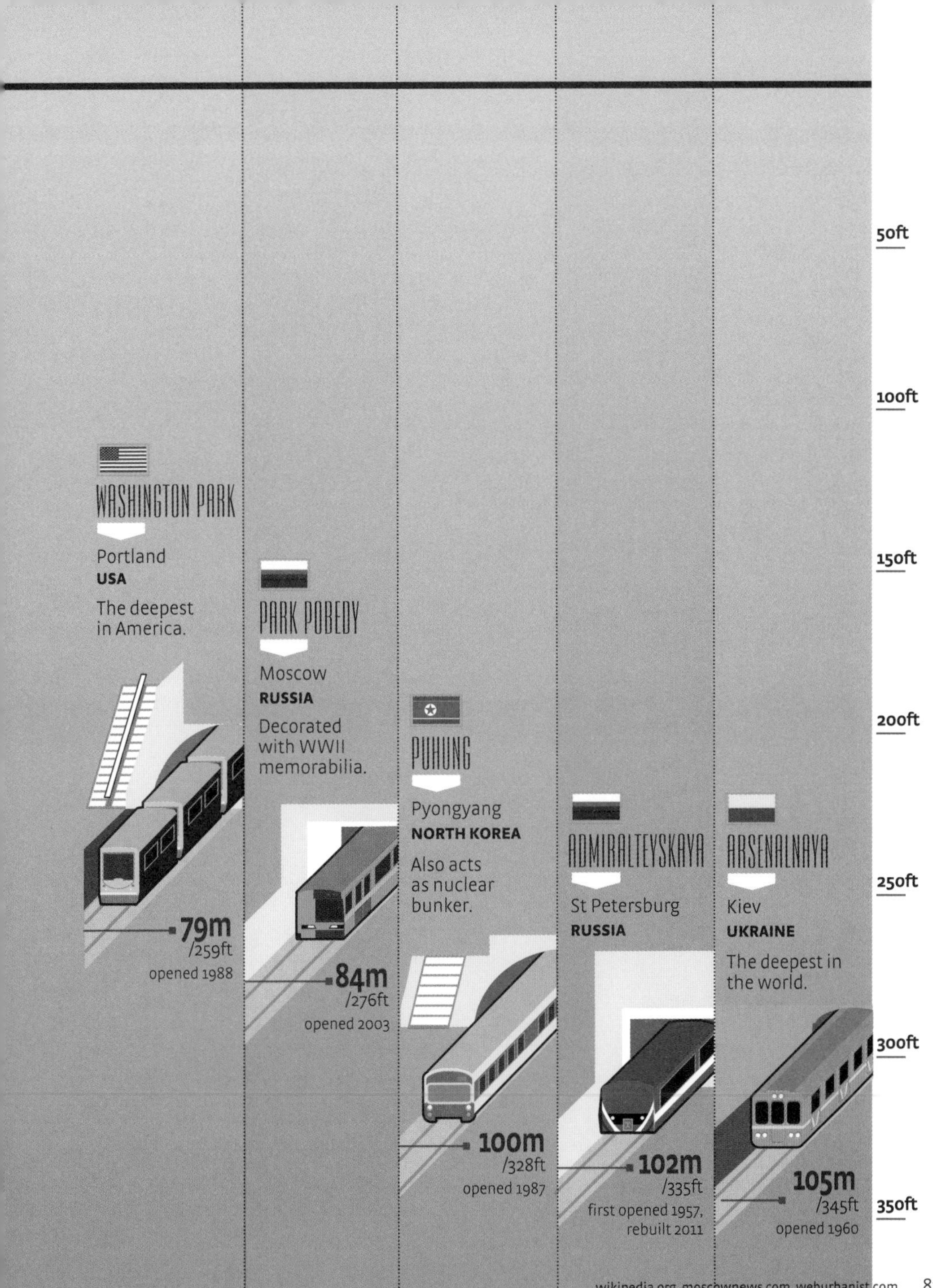

50ft
100ft
150ft
200ft
250ft
300ft
350ft
WASHINGTON PARK
Portland
USA
The deepest in America.
79m /259ft
opened 1988
PARK POBEDY
Moscow
RUSSIA
Decorated with WWII memorabilia.
84m /276ft
opened 2003
PUHUNG
Pyongyang
NORTH KOREA
Also acts as nuclear bunker.
100m /328ft
opened 1987
ADMIRALTEYSKAYA
St Petersburg
RUSSIA
102m /335ft
first opened 1957, rebuilt 2011
ARSENALNAYA
Kiev
UKRAINE
The deepest in the world.
105m /345ft
opened 1960

wikipedia.org, moscownews.com, weburbanist.com

SIX DEGREES OF SEPARATION: **STEPHEN HAWKING**

Dr Stephen Hawking is arguably the most famous physicist in the world. An international best-selling author whose work explains incredibly complicated physics in easy terms, he has produced some of the most profound and original thinking on black holes, time, space and relativity since Albert Einstein. It's little wonder that he's at most six steps away from these great men and women.

1

Winston Churchill (1874–1965, Eng) worked at the Colonial Office in 1921 and employed

T.E. Lawrence (1888–1935, Eng) as an aide, who before that had worked with

Faisal I of Iraq (1883–1933, Iraq) during WWI in the Middle East, where an old friend

Philip Graves (1876–1953 Eng) worked as a journalist; his half-brother

Robert Graves (1895–1985, Eng) the poet, in the 1950s hosted the family of

2

Diophantus (AD201–285) the ancient Greek 'father of algebra' wrote *Arithmetica*, and in the margin of his copy

Pierre de Fermat (1601–65, Fra) wrote his *Last Theorem* in 1637, which was compared with the later work on circles and curves by

Jakob Steiner (1796–1863, Ger), whose triple system was expanded upon by

Thomas Kirkman (1806–95, Eng) a mathematician whose 'schoolgirl problem' inspired a book by

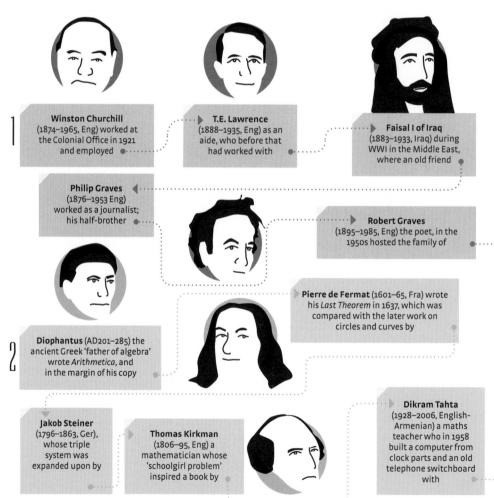

Dikram Tahta (1928–2006, English-Armenian) a maths teacher who in 1958 built a computer from clock parts and an old telephone switchboard with

3 **William Barton Rogers** (1804–82, US), the founder of MIT was the founding president of the American Association for the Advancement of Science (1848), whose first female president

Margaret Burbidge (1919, Eng) was an award-winning astrophysicist who, with her husband

Geoffrey Burbidge (1925–2010, Eng) developed the B^2FH theory, along with

Fred Hoyle (1915–2001, Eng) a Cambridge astrophysicist whose work rejected the Big Bang Theory, and who was supported by

Dennis William Sciama (1926–99, Eng) the 'father of cosmology' who was supervisor at Cambridge University to

Leonard Mlodinow (1954) US physicist, writer on *Star Trek: The Next Generation* and co-author of *A Briefer History of Time* (2005) with

STEPHEN HAWKING

Richard Feynman (1918–88, US), who became an influence of the work of

4 **Albert Einstein**'s (1879–1955) theory of relativity (1905) was the subject of the book and TV series titled *Cosmos* (1980), written by

Carl Sagan (1934–96, US) the cosmologist who employed as a researcher for the book and TV series

Kip Thorne (1940) an American theoretical physicist who worked on black hole theory with

Igor Dmitriyevich Novikov (1935) a Russian cosmologist who also worked on it with

Don Page (1950) a Canadian theoretical physicist who has co-authored journals with

5 **Max Born** (1882–1970), German physicist, mathematician and instrumental in the development of quantum physics, was PhD supervisor to

J. Robert Oppenheimer (1904–67), American father of the atom bomb, who hired

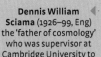

Hans Bethe (1906–2005, Ger) to work on the Manhattan Project where he calculated the bomb's explosive yield along with

WHAT'S **OUT THERE**

Or not, as the case may be, given funding constraints. Here are the space ships that either have been or are intended to go into space with people on board, plus the space stations they are designed to visit.

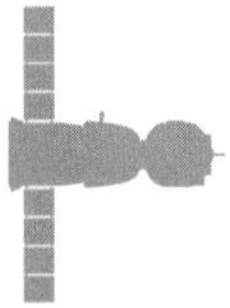

Soyuz Soviet Union-built spacecraft, first unmanned launch 1966, first manned launch 1967, Soyuz ACTS (Advanced Crew Transportation System) launched in 2014.

ATV Automated Transfer Vehicle, European Space Agency funded, unmanned supply space vehicle, 5 launched 2008–2014 to service the International Space Station.

Dragon US-built, first commercially launched spacecraft, launched 2010, unmanned, used to carry cargo to the international space centre. There are plans for a manned version.

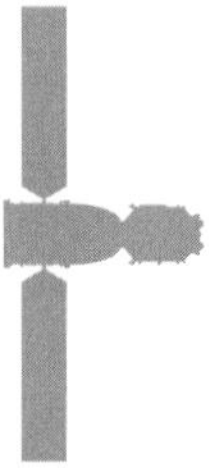

Shenzhou First Chinese-built spacecraft, first unmanned launch 1999, first manned launch 2003, used to reach Chinese space station Tiangong-1.

SpaceShipOne US-built, privately funded sub-orbital air-launched space plane completed first flight in 2003, retired in 2004 and put on public display.

Apollo series Spacecraft developed 1960 to take Americans into space, first unmanned launch 1961, first manned 1968, moon landing 1969, final flight 1975.

Mercury First announced 1958 to take Americans into space, first unmanned sub-orbital flight 1959, first manned sub-orbital flight 1961, last flight 1963, replaced by Apollo.

Vostok Soviet Union-built and launched spaceship designed as a camera/spy ship and to take man into space; first unmanned flight into sub-orbit 1960, first manned sub-orbital flight 1961 (Yuri Gagarin). Retired 1963.

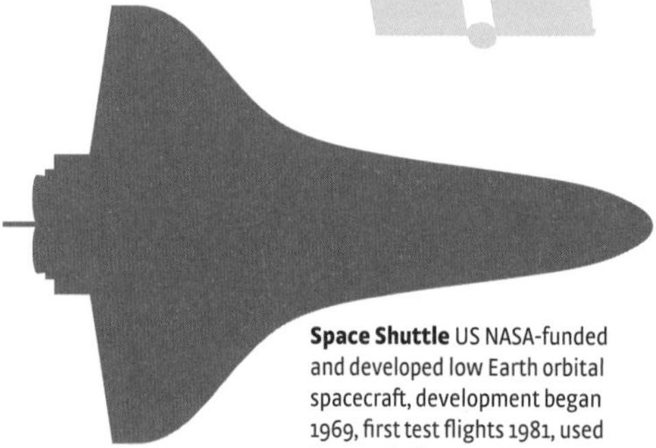

Space Shuttle US NASA-funded and developed low Earth orbital spacecraft, development began 1969, first test flights 1981, used for 135 missions, until 2011.

Genesis I Privately funded American unmanned space habitat, launched 2006 as research into creation of a future modular space station, life expectancy 12 years.

Genesis II Second part of privately funded American plan to launch independent space station, unmanned, launched 2007 to gather data from Genesis I, life expectancy 12 years.

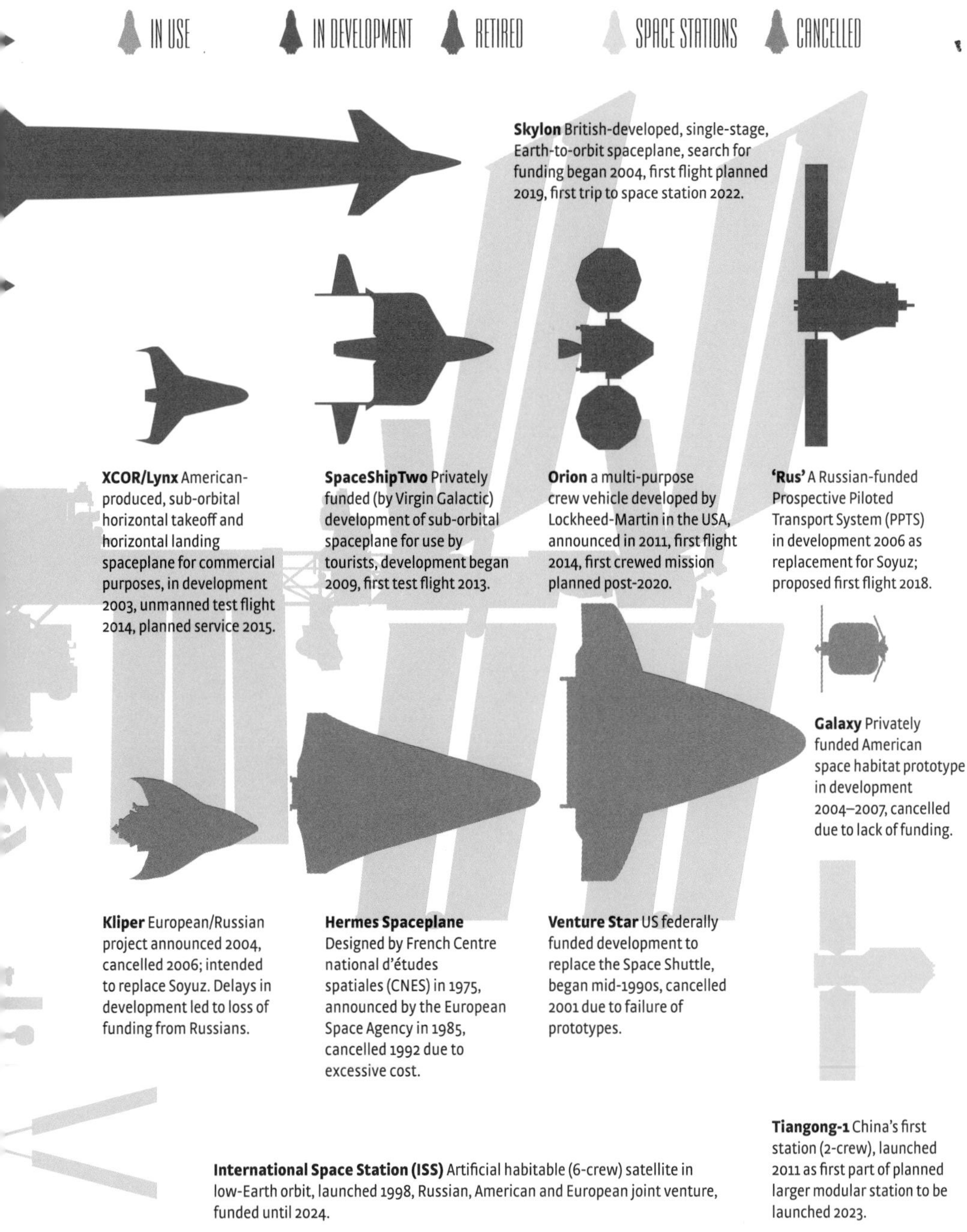
IN USE
IN DEVELOPMENT
RETIRED
SPACE STATIONS
CANCELLED
Skylon British-developed, single-stage, Earth-to-orbit spaceplane, search for funding began 2004, first flight planned 2019, first trip to space station 2022.
XCOR/Lynx American-produced, sub-orbital horizontal takeoff and horizontal landing spaceplane for commercial purposes, in development 2003, unmanned test flight 2014, planned service 2015.
SpaceShipTwo Privately funded (by Virgin Galactic) development of sub-orbital spaceplane for use by tourists, development began 2009, first test flight 2013.
Orion a multi-purpose crew vehicle developed by Lockheed-Martin in the USA, announced in 2011, first flight 2014, first crewed mission planned post-2020.
'Rus' A Russian-funded Prospective Piloted Transport System (PPTS) in development 2006 as replacement for Soyuz; proposed first flight 2018.
Galaxy Privately funded American space habitat prototype in development 2004–2007, cancelled due to lack of funding.
Kliper European/Russian project announced 2004, cancelled 2006; intended to replace Soyuz. Delays in development led to loss of funding from Russians.
Hermes Spaceplane Designed by French Centre national d'études spatiales (CNES) in 1975, announced by the European Space Agency in 1985, cancelled 1992 due to excessive cost.
Venture Star US federally funded development to replace the Space Shuttle, began mid-1990s, cancelled 2001 due to failure of prototypes.
Tiangong-1 China's first station (2-crew), launched 2011 as first part of planned larger modular station to be launched 2023.
International Space Station (ISS) Artificial habitable (6-crew) satellite in low-Earth orbit, launched 1998, Russian, American and European joint venture, funded until 2024.

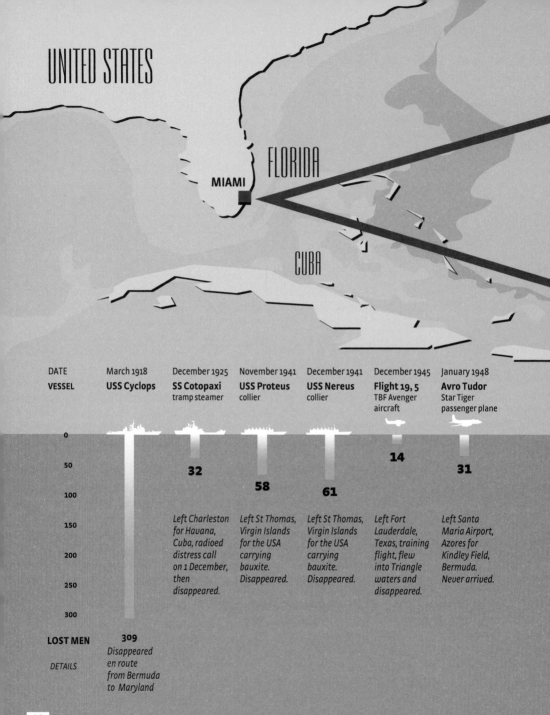

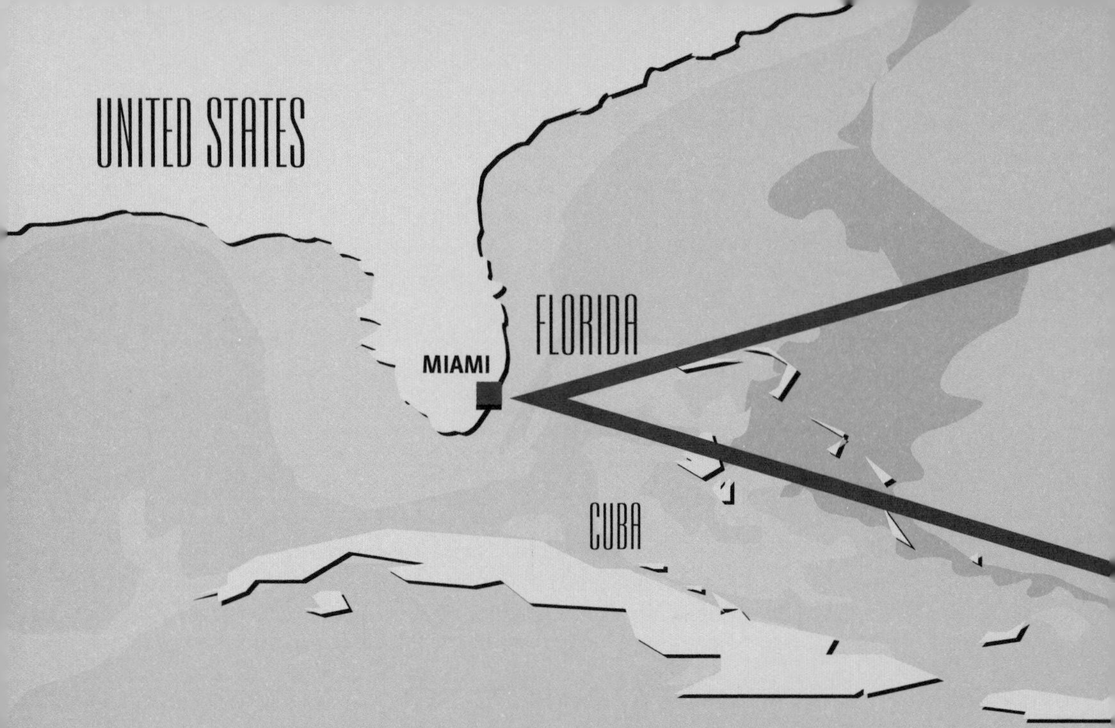

DATE	VESSEL	LOST MEN	DETAILS
March 1918	**USS Cyclops**	309	*Disappeared en route from Bermuda to Maryland*
December 1925	**SS Cotopaxi** tramp steamer	32	*Left Charleston for Havana, Cuba, radioed distress call on 1 December, then disappeared.*
November 1941	**USS Proteus** collier	58	*Left St Thomas, Virgin Islands for the USA carrying bauxite. Disappeared.*
December 1941	**USS Nereus** collier	61	*Left St Thomas, Virgin Islands for the USA carrying bauxite. Disappeared.*
December 1945	**Flight 19, 5** TBF Avenger aircraft	14	*Left Fort Lauderdale, Texas, training flight, flew into Triangle waters and disappeared.*
January 1948	**Avro Tudor** Star Tiger passenger plane	31	*Left Santa Maria Airport, Azores for Kindley Field, Bermuda. Never arrived.*

LOST IN THE **BERMUDA TRIANGLE**

In 1964 Vincent Gaddis, writing in Argosy *magazine, identified an area of the mid-Atlantic as being the Bermuda Triangle. From Bermuda to San Juan, Puerto Rico and Miami, Florida, the 'triangle', Gaddis said, made strange things happen to ships and airplanes. They disappear, mostly. Although debunked and derided as an urban myth, the roll call of vessels lost in the Triangle is long, and the number of missing persons in excess of 800.*

SHIPS ASSUMED TO SAIL THROUGH THE TRIANGLE

December 1948	February 1963	December 1967	March 1938	December 1946	October 1976	October 1996
Douglas DC-3 NC16002 passenger plane	**SS Marine Sulphur Queen** tanker	**Witchcraft** cabin cruiser	**Anglo Australian** freighter	**City Belle** schooner	**Sylvia L Ossa** ore-carrier	**Intrepid** yacht
39	**39**	**2**	**38**	**10**	**37**	**16**
Left San Juan, Puerto Rico for Miami, Florida and disappeared.	*Left Beaumont, Texas and disappeared four days later near Florida.*	*Less than a mile from the Florida coast, the captain radioed shore asking for assistance, then disappeared.*	*Left Cardiff, Wales for British Columbia, last radio report from the Azores.*	*Found abandoned, out of the Bahamas.*	*Missing at approximately 140 miles west of Bermuda.*	*Disappeared en route from Bermuda to Maryland.*

bermuda-triangle.org, wikipedia.org

MOST ENDANGERED LANGUAGE
(1 speaker alive)

1. Apiaká *(Matto Grosso, Brazil)*
2. Diahói *(Southern Amazonas, Brazil)*
3. Kaixána *(Northwest Amazonas, Brazil)*
4. Chaná *(Paraná, Argentina)*
5. Yahgan *(Navarino Island, Chile)*
6. Taushiro *(Northern Peru)*
7. Tinigua *(Guayabero river, Columbia)*
8. Pémono *(Upper Majagua, Venezuela)*
9. Patwin *(Northern California, USA)*
10. Wintu-Nomlaki *(Northern California, USA)*
11. Tolowa *(California/Oregon border, USA)*
12. Bishuo *(Cameroon)*
13. Bikya *(Cameroon)*
14. Dampelas *(Sulawesi, Indonesia)*
15. Pazeh *(Taiwan)*
16. Volow *(Motalava Island, Vanuatu)*
17. Yarawi *(Papua New Guinea)*
18. Laua *(Papua New Guinea)*

ENDANGERED EUROPEAN LANGUAGES
(Number of estimated speakers)

19	Vilamovian	70
20	Karaim	60
21	Votic	20
22	Ingrian	200
23	Ter Saami	10
24	Saterlandic	1000
25	Tsakonian	300
26	Gardiol	340
27	Faetar	600
28	Töitschu	200
29	Cimbrian	400
30	Mòcheno	1000
31	Arbanasi	500
32	Istro-Romanian	300
33	Pite Saami	20
34	Gagauz	400
35	Hértevin	1000
36	Bats	500

ENDANGERED LANGUAGES OF THE UK AND IRELAND
(Number of estimated speakers)

37	Irish Gaelic	77,185	*(in Republic of Ireland – daily speakers, outside of the education system)*
38	Scottish Gaelic	58,000	
39	Guernsey French	1,300	
40	Jersey French	2,000	
41	Cornish	>500	*Extinct in 19th century – since revived, with now over 500 speakers*
42	Manx	100–200	*Extinct in 1970s – since revived, now between 100–200 speakers*
43	Alderney French	–	*Extinct – last native speakers died around 1960*

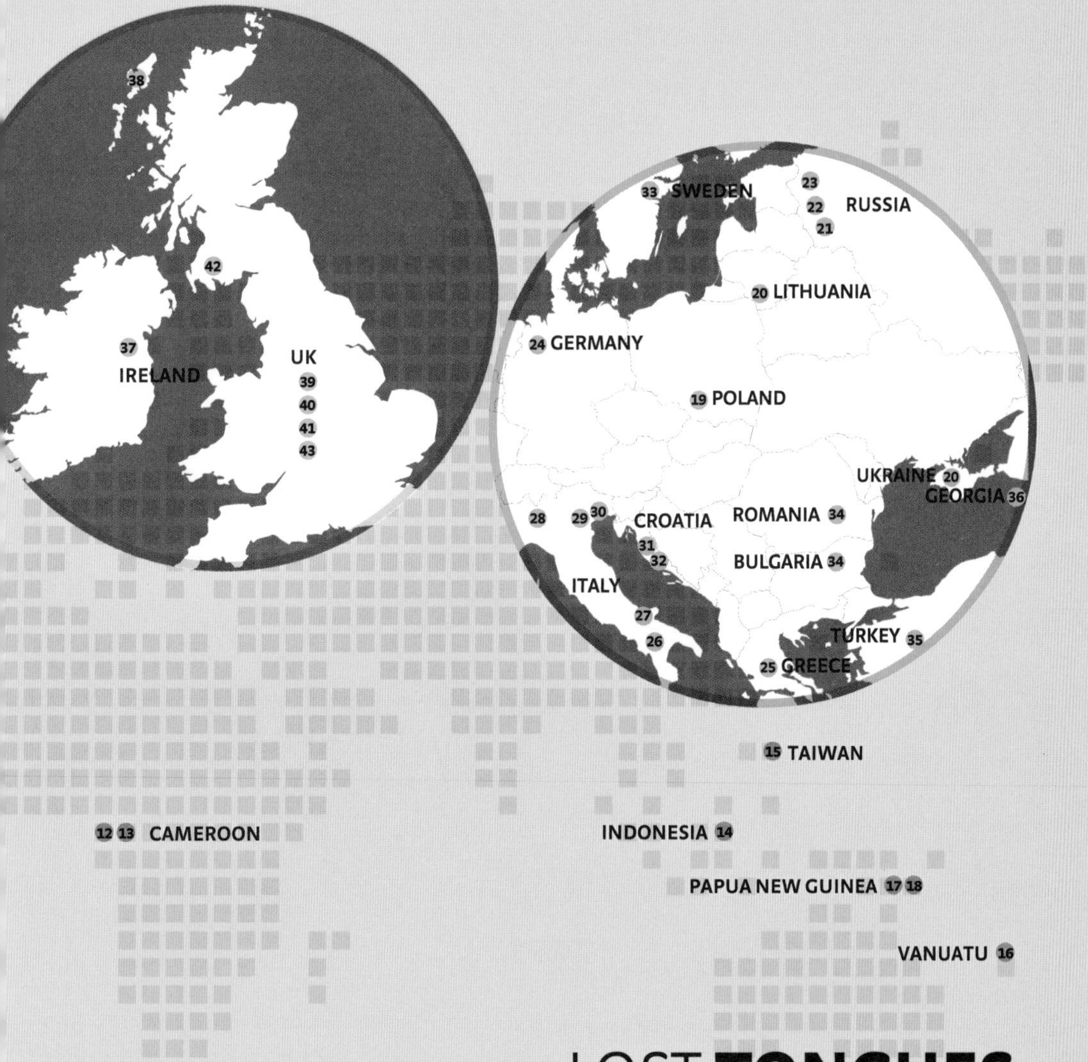

LOST **TONGUES**

As remote cultures are uncovered and integrated with the wider world, so languages become dominated by those of colonizers and conquerors. As native speakers die and their descendants grow up speaking mainstream languages, so mother tongues of indigenous people also die. These are UNESCO's most endangered languages across the world.

bbc.co.uk, thisiscornwall.co.uk, scotslanguage.com, unesco.org

KNOW YOUR **OLIGARCH**

Emerging from the ashes of the Soviet empire in the early 1990s, Russia's oligarchs gained huge personal wealth as they built businesses that dominated the eastern bloc. Their control of essential mineral and fuel resources have enabled them to expand into the West, where they have taken their place among the truly wealthy capitalist elite. Here are the top twenty oligarchs, their wealth and interests outside of business.

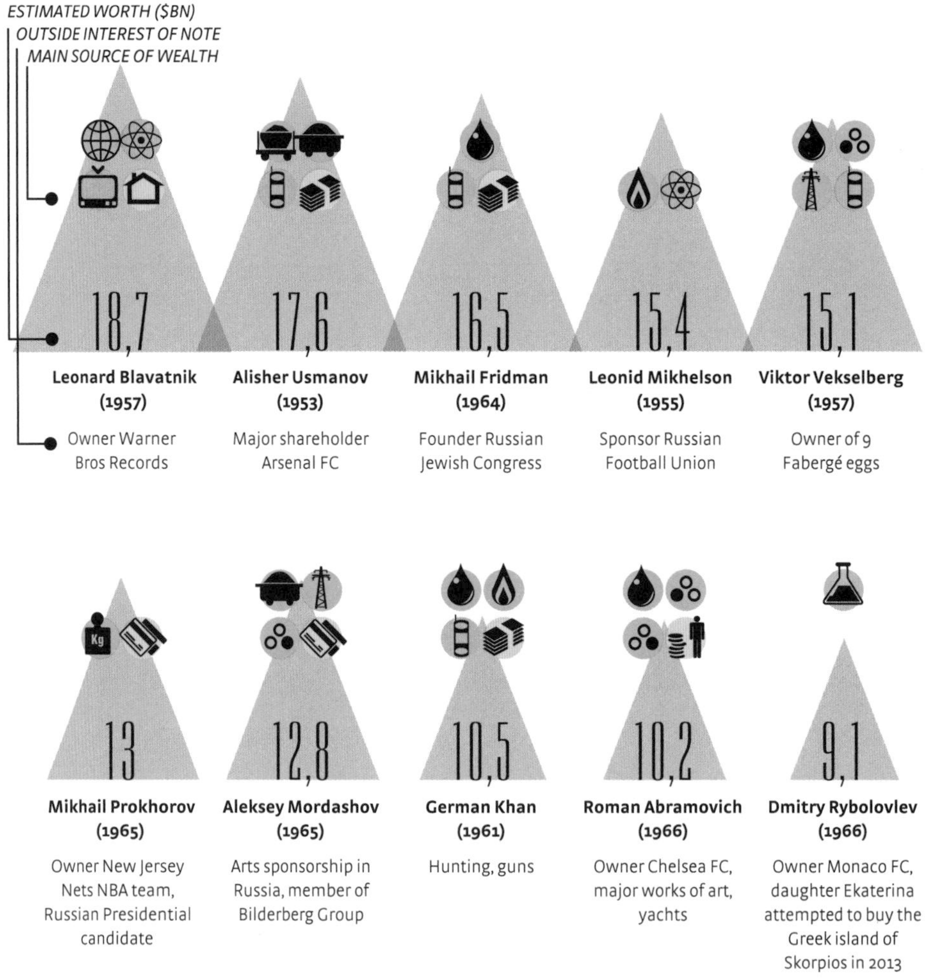

MAIN SOURCE OF WEALTH

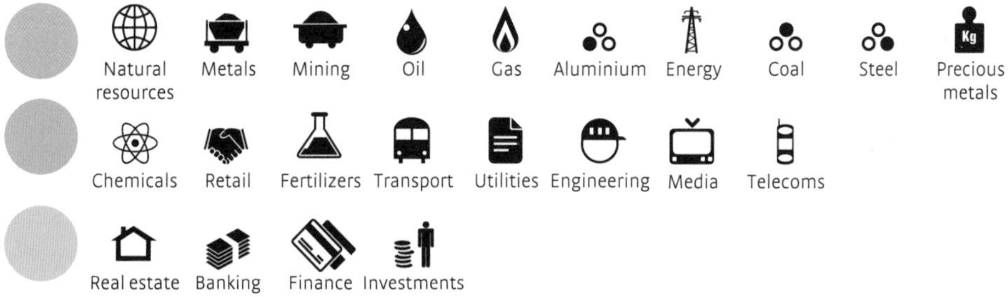

Vagit Alekperov (1963)

Three Russian TV companies, a newspaper and magazine

Andrey Melnichenko (1972)

Owner motor yacht 'A', cost $300m

Vladimir Potanin (1961)

On the board of trustees of the Guggenheim Foundation (NY)

Vladimir Lisin (1956)

President of Russian Shooting Union

Gennady Timchenko (1952)

Part-owner Jokerit, a Finnish ice hockey team; sponsor of chess tournament Alekhine Memorial

Iskander Makhmudov (1963)

Owner, the ice arena in Sochi, Russian business magazines

Oleg Deripaska (1968)

On the board of trustees of the Bolshoi Ballet, established Volnoe Delo

Sergei Galitsky (1967)

Owner FC Krasnodar

Alexey Kuzmichev (1962)

Founder of the Babylon Project to preserve ancient artefacts of Iraq

Andrei Skoch (1966)

Philanthropy

From the tiniest living thing to the largest organism is a huge distance. The smallest creature can't be seen without a microscope while it would take a human more than 30 minutes to run the length of the largest. Here are the world's dozen creatures great and small for comparison.

RETROVIRUS

SIZE: 80nm
some scientists do not regard viruses as living organisms because they do not have cellular structure

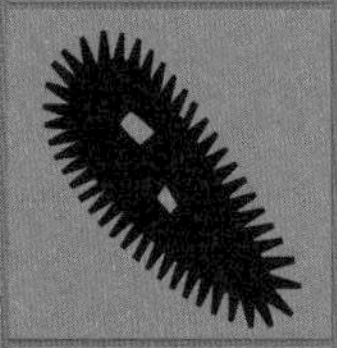

SINGLE-CELLED ARCHAEA

SIZE: 200nm
one-fifth the size of a typical bacterium

MYCOPLASMA GENITALIUM

SIZE: 200–300nm
the smallest bacterium

DICYEMIDA

(or Rhombozoa)

SIZE: 500μm
the multicelled life form, a parasite, with fewest cells: 20

NEMATODE WORM

SIZE: 80μm long, 5μm wide
the animal with the fewest cells; less than 1000

STYGOTANTULUS STOCKI

SIZE: 100μm
the smallest invertebrate, a crustacean

FROG

(Paedophryne amauensis)

SIZE: 8mm
the smallest invertebrate, from Papua New Guinea

ALL CREATURES GREAT AND SMALL

AFRICAN BUSH ELEPHANT

(Loxodonta africana)

SIZE: 12 tonnes
the largest land animal alive

GIRAFFATITAN

SIZE: 37 tonnes
the largest animal ever to have lived, a sauropod dinosaur

SIZE: 190 tonnes
the largest animal ever to have lived that lives still

BLUE WHALE

(Balaenoptera musculus)

GIANT SEQUOIA TREE

(Sequoiadendron giganteum)

SIZE: 87m/285ft the largest tree still growing, named Diamond, in Atwell Mill Grove, Silver City, CA, USA

SIZE: $9km^2/5mi^2$; 600+ tonnes
the largest living organism, over 2400 years old, in the Blue Mountains of eastern Oregon, USA

HONEY FUNGUS
ARMILLARIA OSTOYAE

sfgate.com, wikipedia.org

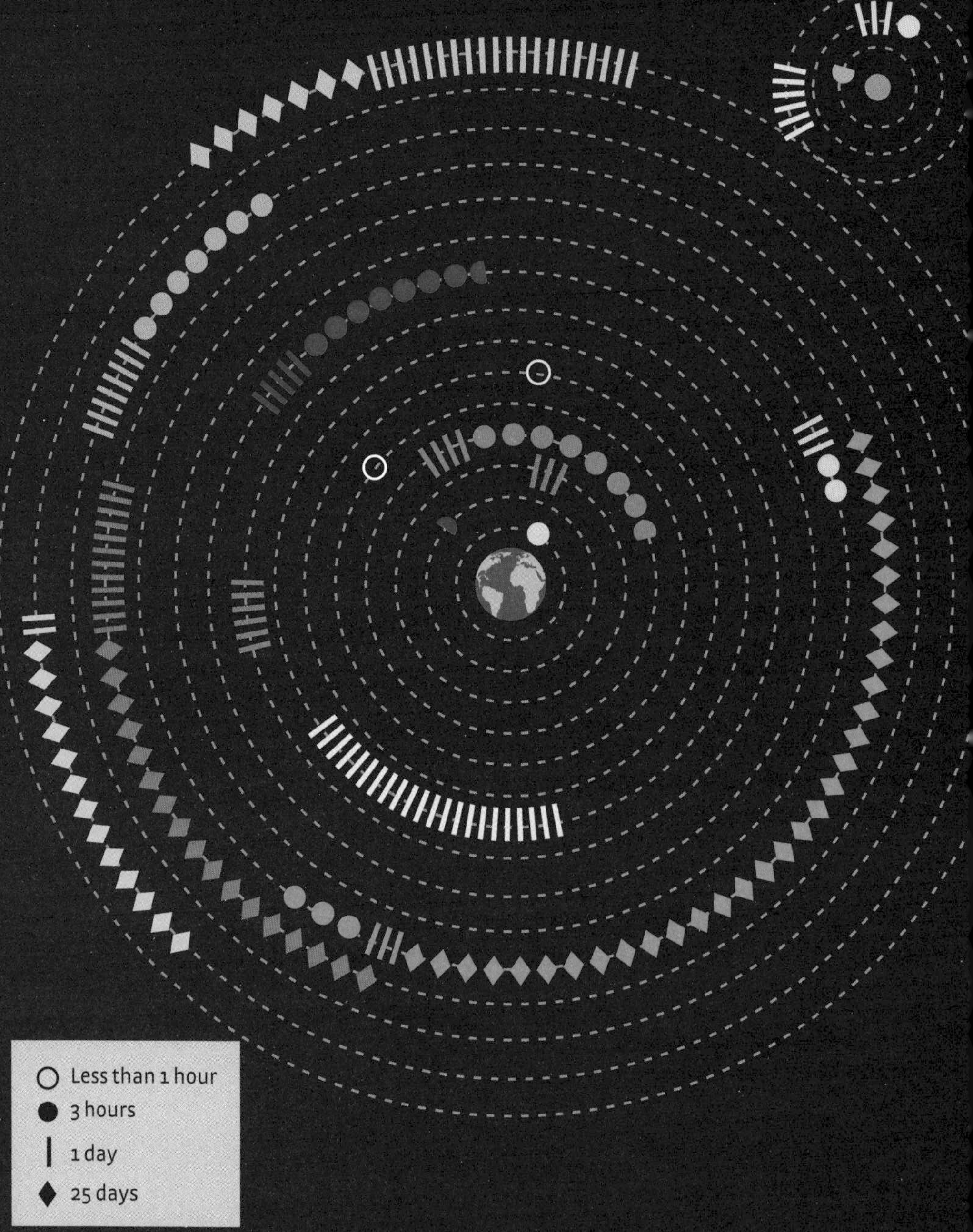

○	Less than 1 hour
●	3 hours
│	1 day
◆	25 days

BOLDLY GOING

As of November 6, 2013, a total of 536 people had traveled into space, with three making only sub-orbital flights and the others getting into Earth orbit. Of that number 24 have gone beyond low Earth orbit and 12 have walked on the Moon. Laika the Russian dog got into orbit before anyone else from Earth. Here are the landmark excursions into space.

1957 Laika (dog)
circuited the Earth

1961 Yuri Gagarin
first man to orbit the Earth

1963 Valentina Tereshkova
first woman in space

1963 Valery Bykovsky
took the longest solo flight into space

1965 Alexey Leonov
performed first spacewalk

1965 Edward H. White II
spacewalk, first use a hand-held manoeuvring unit for 20 secs

1966 Veterok and Ugolyok

1968 Frank Borman, James Lovell and William Anders
Apollo 8, first manned rocket to leave Earth's orbit

1969 Neil Armstrong and Buzz Aldrin
Apollo 11, the first mission to land men on the moon

1970 Jim Lovell, Fred Haise and John Swigert
Apollo 13, first humans 400,171km from Earth

1972 Eugene Cernan and Harrison Schmitt
spent longest stay on the lunar surface

1972 Ronald Evans
had the longest stay in lunar orbit aboard Apollo 17

1988–1998 Anatoly Solovyev
record number of spacewalks (16) and longest cumulative time spent spacewalking

1988–2005 Sergei Krikalev
spent more time in space (cumulative) than anyone else

1994–1995 Valeri Polyakov
spent longest time in space aboard the Mir space station

1998 John Glenn
the oldest man to go into space, aged 77

2002–2007 Peggy Whitson
most time in space for a female

2006–2007 Sunita Williams
the longest single flight into space for a woman

news.bbc.co.uk, space today.org, wikipedia.org

LIVING UNDER A **VOLCANO**

The people fossilized in Pompeii by the eruption of Vesuvius in AD 79 (estimated 3300 killed) is a chilling warning to those people who reside well within lava range of volcanoes. Surprisingly there are a lot of them, as this graphic shows.

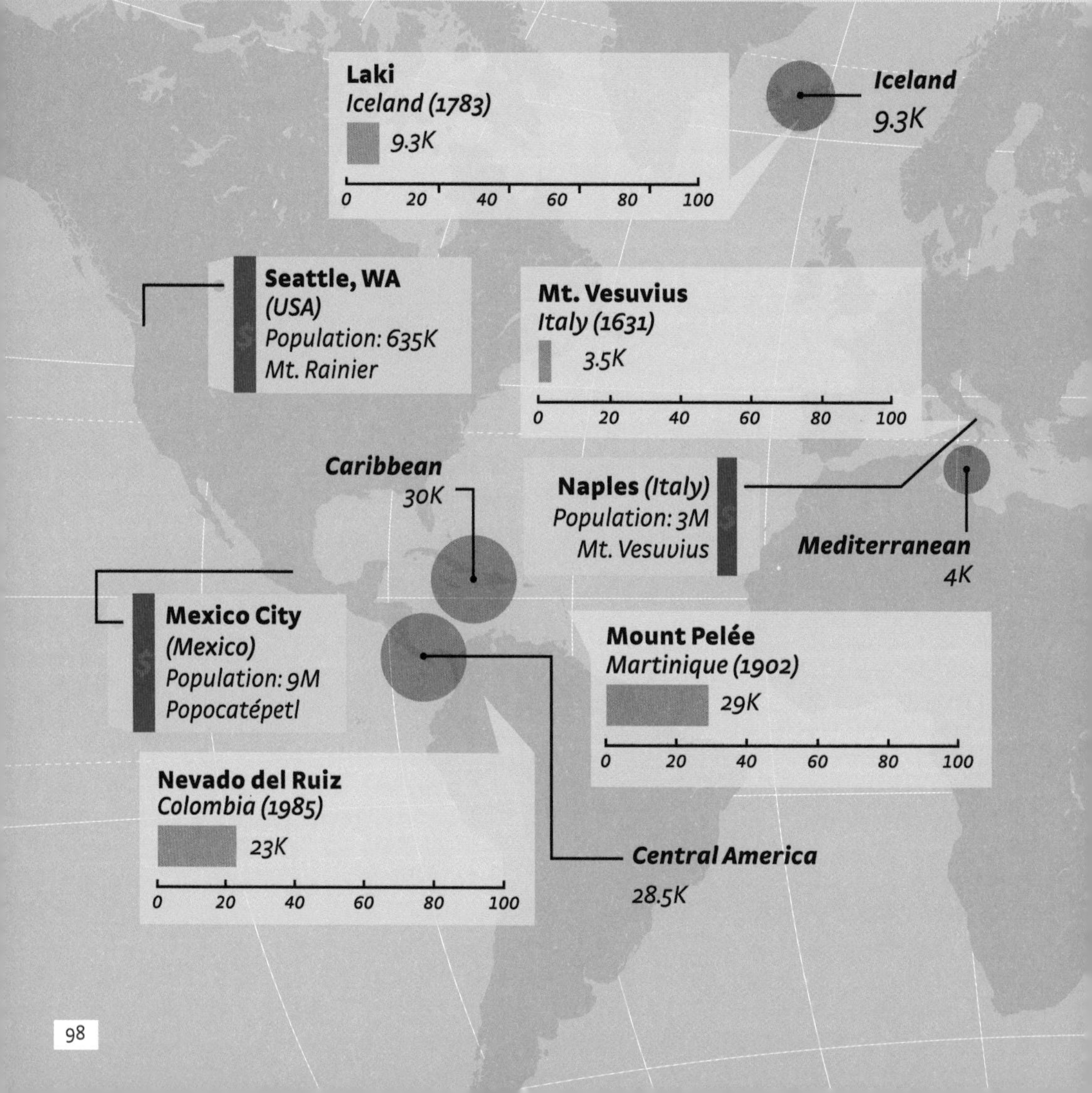

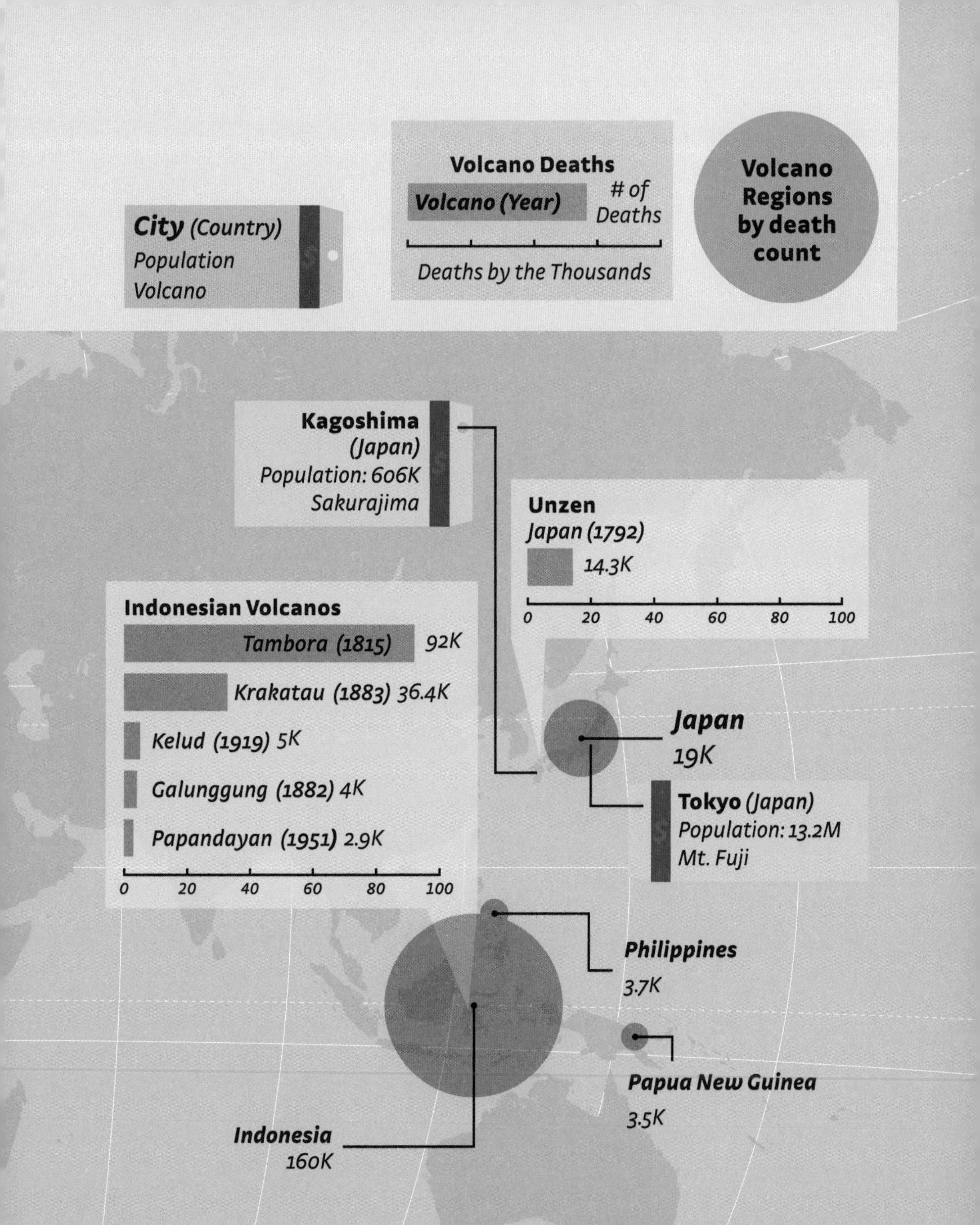

volcano.oregonstate.edu, Blong, R.J., 1984, Volcanic Hazards: A Sourcebook on the Effects of Eruptions: Orlando, Florida, Academic Press, volcanodiscovery.com, wikipedia.org

LIGHT UP THE WORLD

ght pollution is the unwanted effect of artificial light, which can block out the stars nd disrupt natural biological rhythms in humans and animals. In March 2013, Hong ong was named most light-polluted city on Earth. Among the least light-polluted ies are the protected regions designated as International Dark Sky Parks (IDSPs) and eserves (IDSRs). The Bortle Scale shown here gauges regions for light pollution.

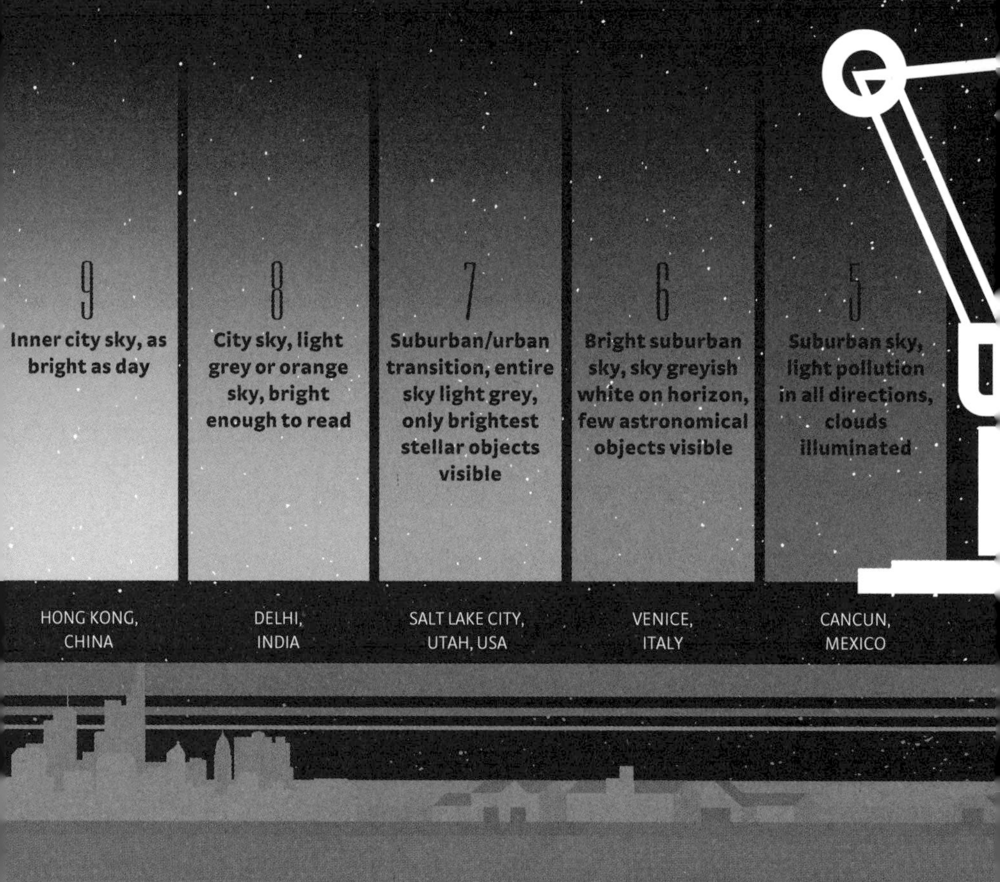

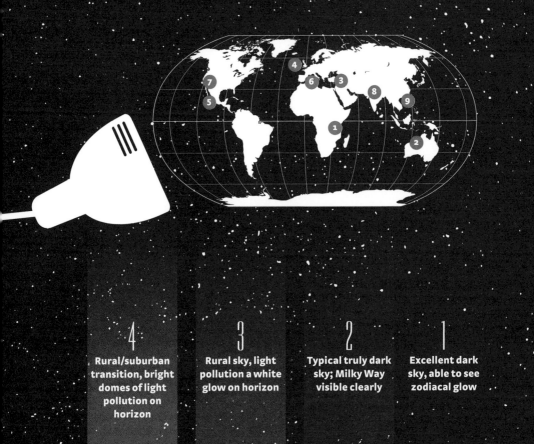

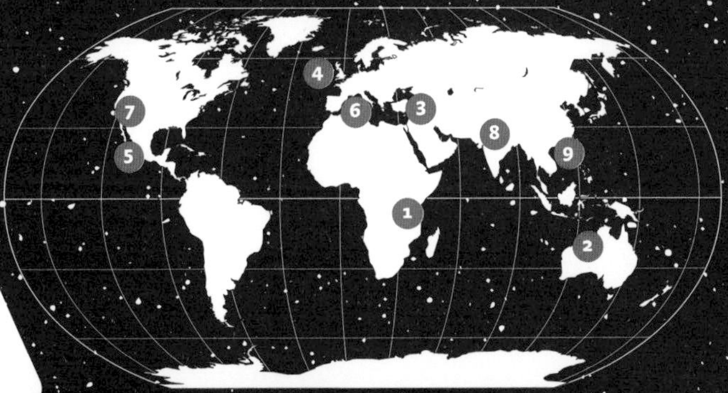

4
Rural/suburban transition, bright domes of light pollution on horizon

TRUSK LOUGH, BALLYBOFEY, COUNTY DONEGAL, IRELAND

3
Rural sky, light pollution a white glow on horizon

ULUDAG NATIONAL PARK, BURSA, TURKEY

2
Typical truly dark sky; Milky Way visible clearly

BUNGLE RANGE, WESTERN AUSTRALIA

1
Excellent dark sky, able to see zodiacal glow

MOUNT KILIMANJARO, TANZANIA, AFRICA

ELEMENTARY **NOMENCLATURE**

The chemical elements named after the people who discovered them or the places where they were first described.

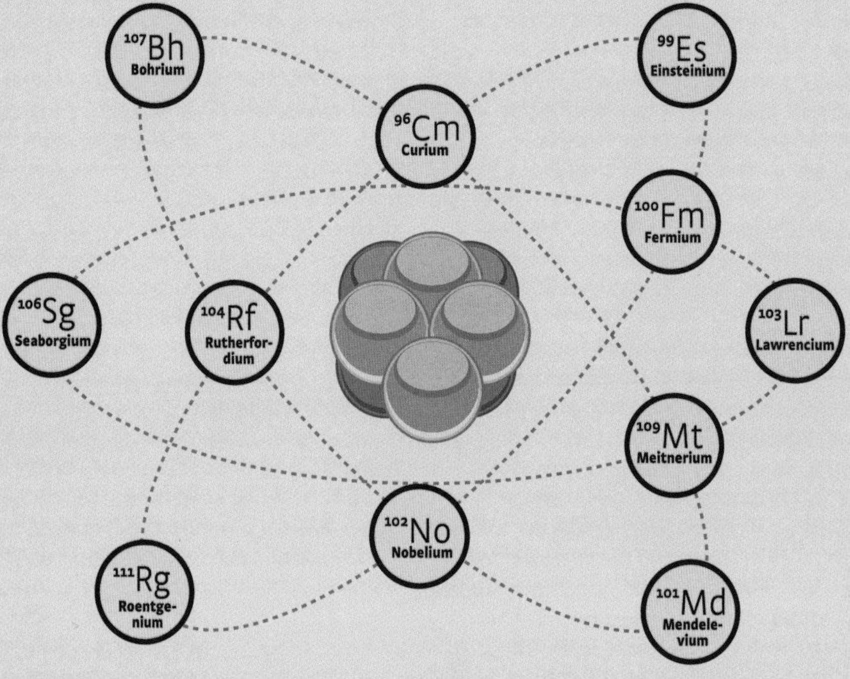

ELEMENTS NAMED AFTER PEOPLE

(element, symbol, atomic number, name origin, date of discovery)

Bh	Bohrium, Bh, 107, Niels Bohr, 1981
Cm	Curium, Cm, 96, Pierre and Marie Curie, 1944
Es	Einsteinium, Es, 99, Albert Einstein, 1952
Fm	Fermium, Fm, 100, Enrico Fermi, 1952
Lr	Lawrencium, Lr, 103, Ernest Lawrence, 1961
Mt	Meitnerium, Mt, 109, Lise Meitner, 1982
Md	Mendelevium, Md, 101, Dmitri Mendeleev, 1955
No	Nobelium, No, 102, Alfred Nobel, 1956
Rg	Roentgenium, Rg, 111, Wilhelm Roentgen, 1994
Rf	Rutherfordium, Rf, 104, Ernest Rutherford, 1964
Sg	Seaborgium, Sg, 106, Glenn T. Seaborg, 1974

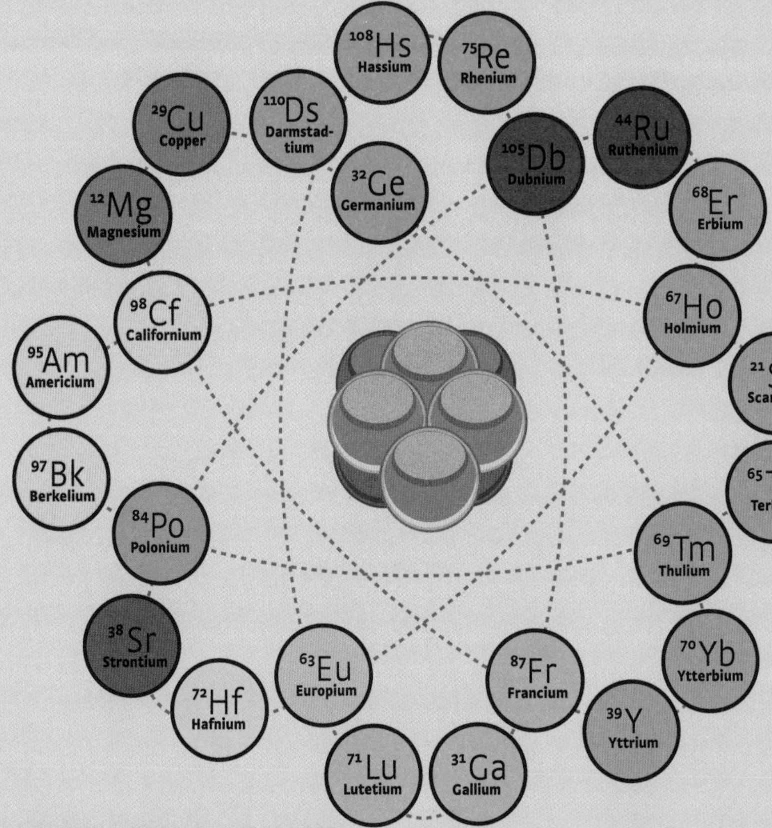

ELEMENTS NAMED AFTER PLACES *(element, symbol, atomic number, name origin, date of discovery)*

Am	Americium, Am, 95, America, 1944	Re	Rhenium, Re, 75, Rhenus (Latin name for the Rhine), 1925	Y	Yttrium, Y, 39, Ytterby in Sweden, 1794
Bk	Berkelium, Bk, 97, Berkeley in California, 1949	Db	Dubnium, Db, 105, Dubna in Russia, 1967–70	Fr	Francium, Fr, 87, France, 1939
Cf	Californium, Cf, 98, state and University of California, 1950	Ru	Ruthenium, Ru, 44, Ruthenia (Latin name for Russia), 1808	Ga	Gallium, Ga, 31, Gallia (Latin name for France), 1875
Cu	Copper, Cu, 29, Cyprus, 5th century BC	Er	Erbium, Er, 68, Ytterby in Sweden, 1843	Lu	Lutetium, Lu, 71, Lutetia (Roman name for Paris), 1907
Mg	Magnesium, Mg, 12, Magnesia (district in Greece), 1755	Ho	Holmium, Ho, 67, Holmia (Latin name for Stockholm), 1878	Eu	Europium, Eu, 63, Europe, 1901
Ds	Darmstadtium, Ds, 110, Darmstadt in Germany, 1994	Sc	Scandium, Sc, 21, Scandia (Latin name for Scandinavia), 1879	Hf	Hafnium, Hf, 72, Hafnia (Latin name for Copenhagen), 1923
Ge	Germanium, Ge, 32, Germany, 1886	Tb	Terbium, Tb, 65, Ytterby in Sweden, 1843	Sr	Strontium, Sr, 38, Strontian in Scotland, 1790
Hs	Hassium, Hs, 108, Hesse in Germany, 1984	Tm	Thulium, Tm, 69, Thule (ancient name for Scandinavia), 1879	Po	Polonium, Po, 84, Poland, 1898
		Yb	Ytterbium, Yb, 70, Ytterby in Sweden, 1878		

WHAT **PLANET** ARE YOU ON?

If you'd been rocketed into space with no map or directions, and with only a very long tape measure in your spacesuit, would you be able to tell which planet you'd landed on from its mean radius alone?

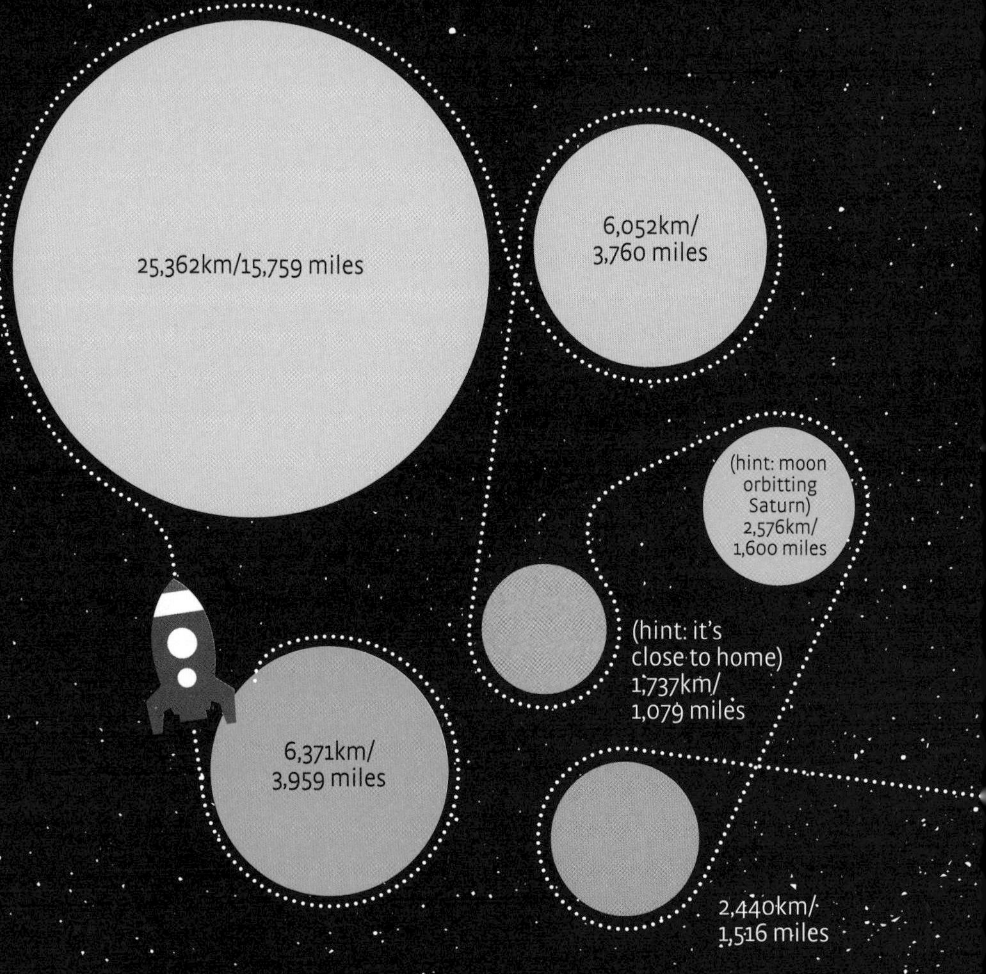

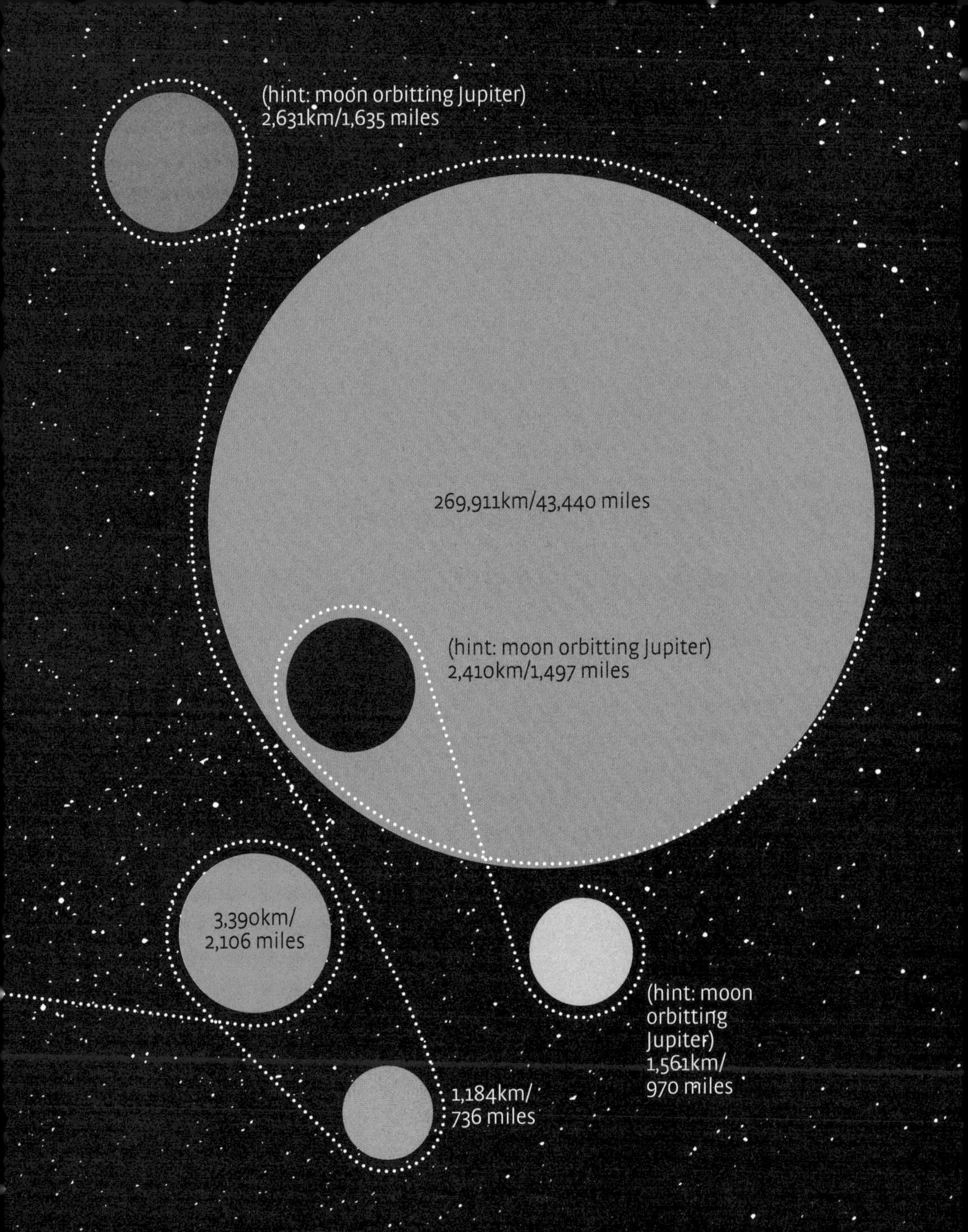
(hint: moon orbitting Jupiter)
2,631km/1,635 miles
269,911km/43,440 miles
(hint: moon orbitting Jupiter)
2,410km/1,497 miles
3,390km/
2,106 miles
(hint: moon
orbitting
Jupiter)
1,561km/
970 miles
1,184km/
736 miles

STAMPING OUT **DISEASES**

Deaths
Blinded
Infected

Despite spending billions of dollars around the world to end the spread of diseases, smallpox was the first and so far the only major infectious disease in humans to be successfully eradicated (in 1980). These are the next top targets for eradication by the World Health Organization.

DISEASE TARGETED BY THE INTERNATIONAL TASKFORCE FOR DISEASE ERADICATION

TENIASIS/ CYSTICERCOSIS
(pork tapeworm)

50M CASES
50,000 DEATHS
PREVALENT IN
Africa, Asia, Latin America

MEASLES

780,000 DEATHS
PREVALENT IN
Africa, South-East Asia, Europe, Eastern Mediterranean, Western Pacific regions

POLIOMYELITIS *(Polio)*

2000 CASES OF PARALYTIC DISEASE
200 DEATHS
PREVALENT IN
Afghanistan, Nigeria, Pakistan

LYMPHATIC FILARIASIS

120 MILLION INFECTED
PREVALENT IN
65% South-East Asia, 30% Africa

RUBELLA *(CRS)*

10,000 BABIES BORN WITH CONDITION
PREVALENT IN
Africa, South-East Asia

DRACUNCULIASIS
(Guinea worm disease)

> 10,000 PERSONS INFECTED
PREVALENT IN Africa

AMERICAN TRYPANOSOMIASIS
(Chagas disease)

10-12M INFECTED

PREVALENT IN
Latin America

TRACHOMA

2.2M CASES

1.2M BLIND

PREVALENT IN
Africa, Latin America, Asia, Australia

RABIES

52,000 DEATHS

PREVALENT IN
Asia, Africa

NEONATAL TETANUS

560,000 DEATHS

PREVALENT IN
Rural Africa, Asia, Latin America

MALARIA

> 50M CASES

> 750,000 DEATHS

PREVALENT IN
Asia, Africa, Latin America

HEPATITIS B

600,000 DEATHS

PREVALENT IN
Sub-Saharan Africa, East Asia

ONCHOCERCIASIS
(river blindness)

37-40M CASES

340,000 LEFT BLIND

PREVALENT IN
Sub-Saharan Africa

PARTIALLY AND POTENTIALLY ERADICABLE DISEASES

who.org, Oxford Companion to Medicine, cartercenter.org

MAY YOUR **GOD** GO WITH YOU

He may have different names, but God is still a powerful figure for believers around the world. Here's how the 7 billion people on Earth divide in terms of religious belief, or non-belief.

31%

CHRISTIANITY

Catholic 50%

Protestant 37%

Orthodox 12%

other 1%

7%

BUDDHIST

50%

in China

16%

NO RELIGION

Atheists, agnostics

GOD IS DEAD

0.2%

JEWISH

41%

in USA

41%

in Israel

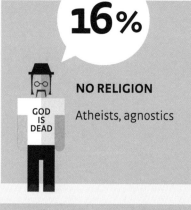

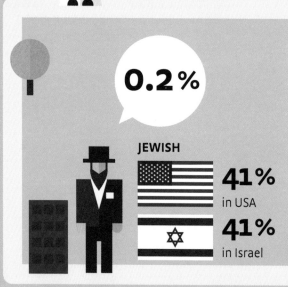

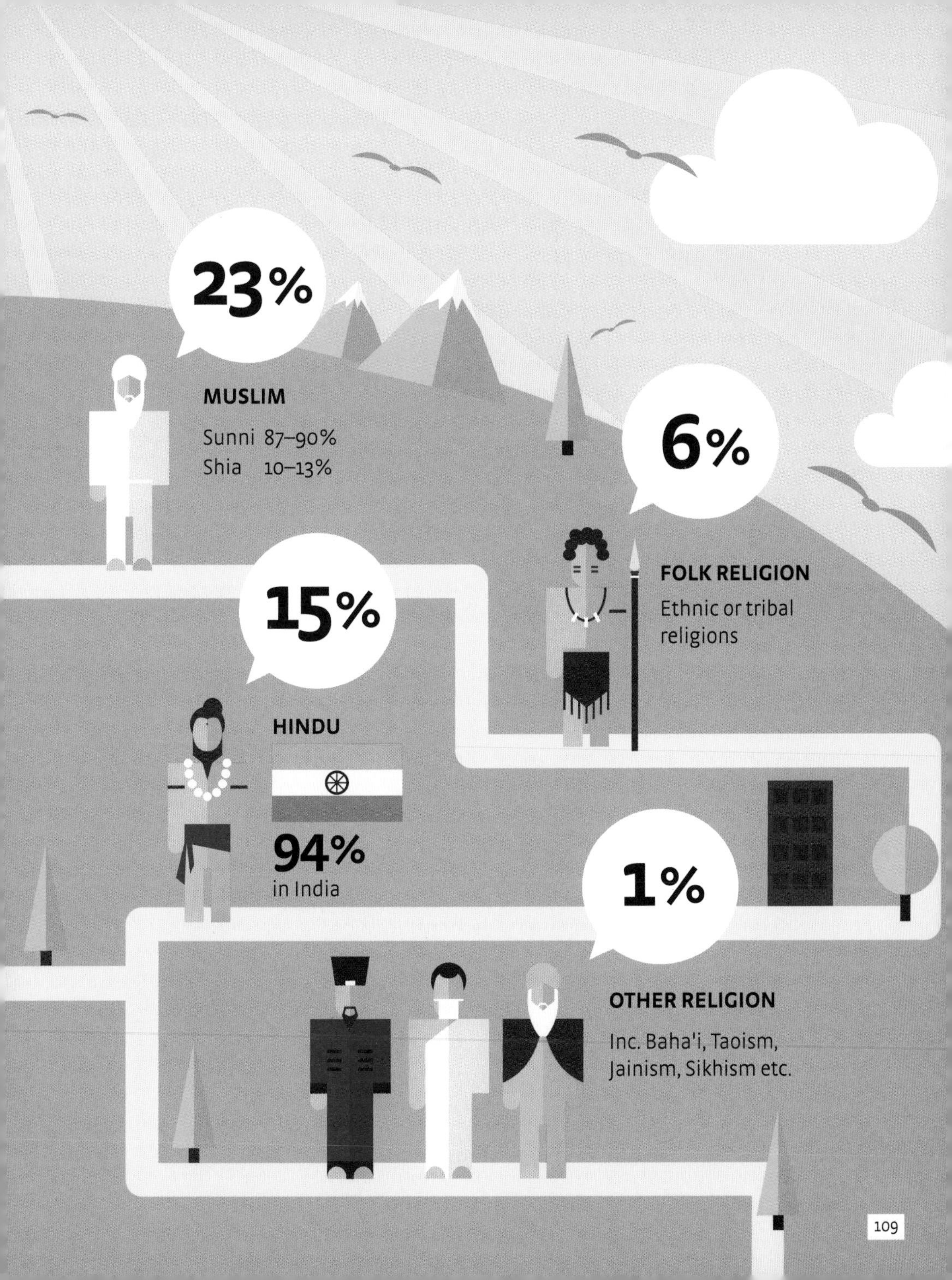
23%
MUSLIM
Sunni 87–90%
Shia 10–13%
6%
FOLK RELIGION
Ethnic or tribal religions
15%
HINDU
94%
in India
1%
OTHER RELIGION
Inc. Baha'i, Taoism, Jainism, Sikhism etc.

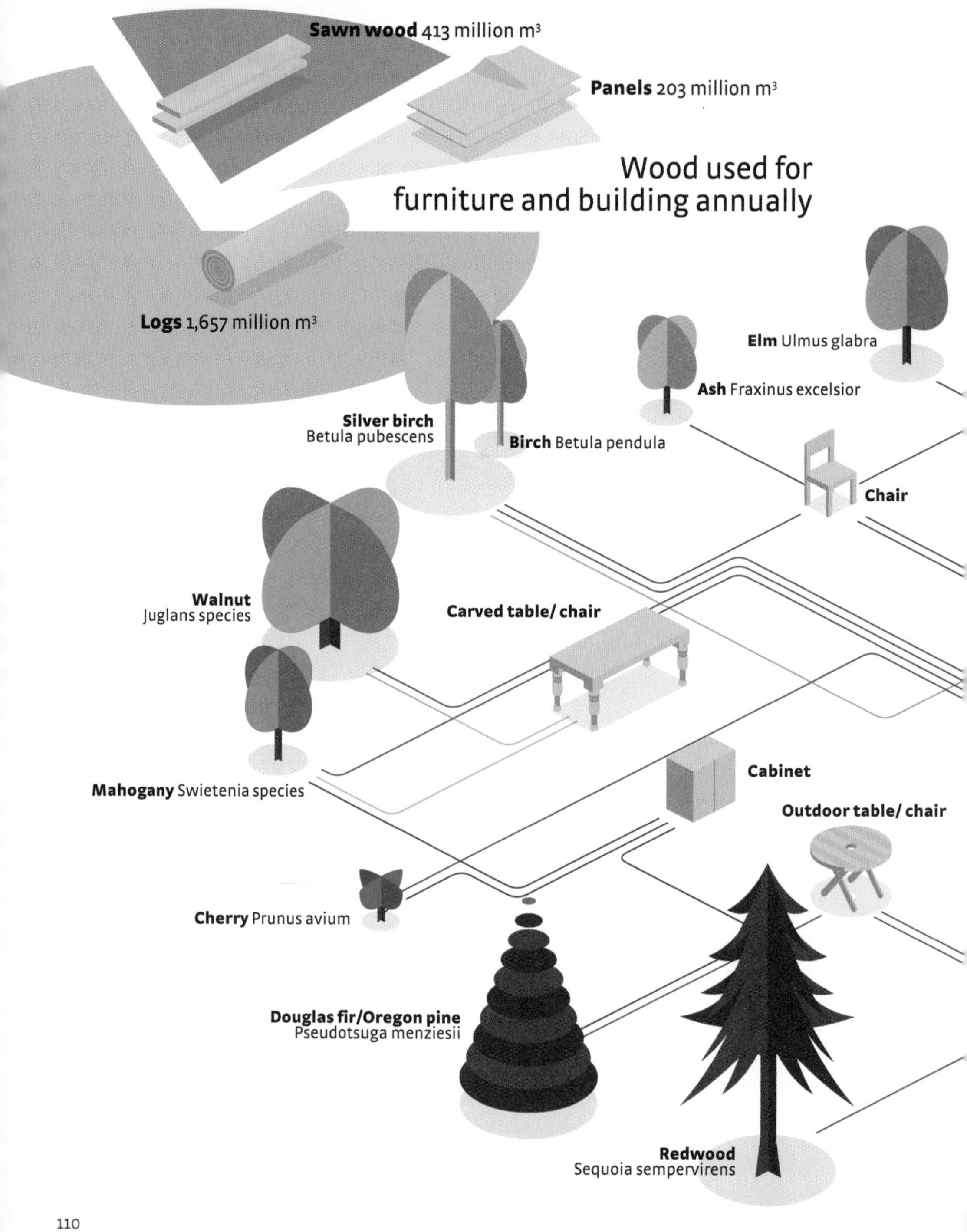
Sawn wood 413 million m³
Panels 203 million m³
Wood used for furniture and building annually
Logs 1,657 million m³
Elm Ulmus glabra
Ash Fraxinus excelsior
Silver birch Betula pubescens
Birch Betula pendula
Chair
Walnut Juglans species
Carved table/ chair
Mahogany Swietenia species
Cabinet
Outdoor table/ chair
Cherry Prunus avium
Douglas fir/Oregon pine Pseudotsuga menziesii
Redwood Sequoia sempervirens

FLATPACK **FORESTS**

The rise in popularity of flatpack furniture over the past two decades in the US and Europe has seen a whole new industry in forestry emerge. Here's how much wood – and what type – gets used annually in the production of everything from tennis racquets to chairs, tables and panelling.

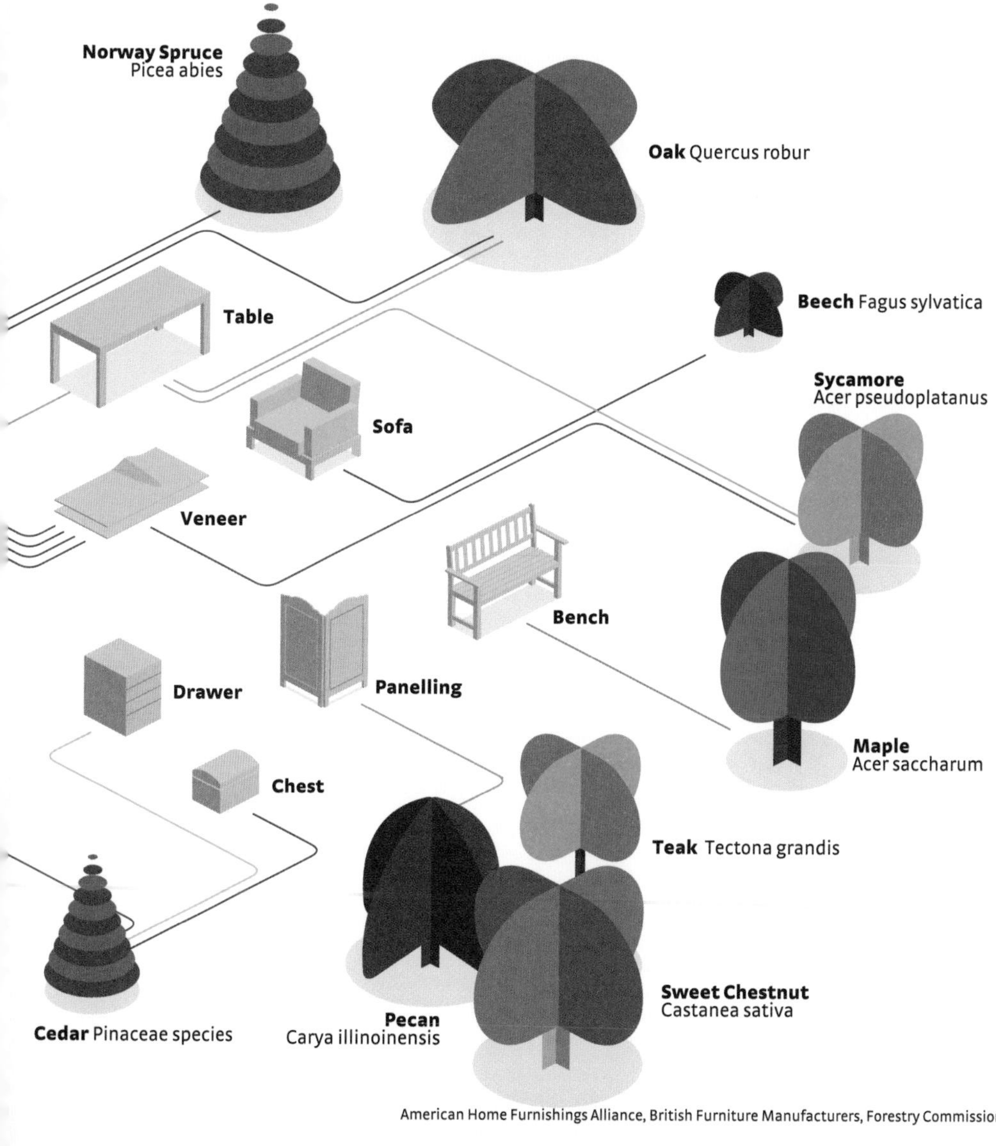

American Home Furnishings Alliance, British Furniture Manufacturers, Forestry Commission

ABACUS TO **TWITTER**

The computer didn't strictly speaking begin with Charles Babbage. He was just one stop on the long ticker tape journey to Twitter:

c. 3000 BC
Abacus counting device invented in the Middle East or China.

1890
Herman Hollerith designs a tabulating machine to process the data for the 1890 US Census. He later founds the Tabulating Machine Company, the forerunner of IBM.

1930
Alan Turing describes the principles behind the modern computer, the 'universal Turing machine'.

1939
German Konrad Zuse designs the Z1, the first mechanical binary programmable computer.

1939
Hewlett-Packard company founded in Palo Alto, California, beginning the area's long association with computing.

1943
Colossus, a large vacuum tube computer, is developed by the British to break the German wartime codes at Bletchley Park.

1944
Harvard University and IBM collaborate to build the Mark 1, the first programmable digital computer. It was 16 meters long, weighed 4,500 kg and was made from 765,000 components.

1946
ENIAC (Electronic Numerical Integrator And Computer) is unveiled. Designed for use at the US Army's Ballistic Research Lab, it is regarded as the first electronic digital computer.

1956
MANIAC at the Los Alamos National Laboratory in New Mexico becomes the first computer to play a full game of chess.

1981
C:\>_
The MS-DOS operating system is introduced on the IBM personal computer.

1982
Elk Cloner, the first major self-spreading personal computer virus, is written by 15-year-old schoolboy Rich Skrenta.

1983
DNS
Domain Name System introduced for the Internet.

1984
Apple introduces the Macintosh computer.

1988
The Morris worm is distributed via the Internet.

1989
www.
Tim Berners-Lee of the particle physics laboratory CERN in Geneva develops the World Wide Web, to help scientists around the world collaborate.

1990
Berners-Lee puts the first website online at http://info.cern.ch

1994
The White House launches its website, www.whitehouse.gov.

2005
First video uploaded to YouTube, 'Me at the Zoo', by co-founder Jawed Karim.

2006
Founder Jack Dorsey posts the first message on Twitter, 'just setting up my twttr'.

2010
WikiLeaks website posts thousands of US State department diplomatic cables online.

2012
B
Number of active users of Facebook passes the billion mark.

1st century BC
Antikythera mechanism is built in Greece to calculate astronomical positions.

Early 17th century
Scottish mathematician John Napier devises logarithms. They provide the basis for the operation of William Oughtred's slide rule.

1640
French mathematician Blaise Pascal builds a mechanical calculating device.

1801
Joseph Marie Jacquard demonstrates his loom for weaving silk; the instructions for the pattern contained on punch cards helped lay the foundations for later computer programming.

1820
Charles Babbage starts work on his Difference Engine calculating machine. Only a demonstration piece with around 2,000 parts is ever assembled.

1830
Babbage develops plans for the Analytical Engine, regarded as the forerunner of the modern computer.

1840
Ada, Countess of Lovelace, former collaborator with Babbage, publishes instructions for the Analytical Engine – effectively the first description of a 'computer program'.

1854
George Boole publishes his Laws of Thought. He devised what would come to be known as Boolean logic, the basis of modern computer science.

1960
Doug Engelbart invents the computer mouse, then known as the X-Y Position Indicator.

1967
IBM team starts to develop the first floppy disk.

1971
Ray Tomlinson sends the first network e-mail. The '@' symbol he chose is still used in email addresses today.

1971
Intel 4004 becomes the first commercially available microprocessor.

Mid-1970s
TCP/IP
Vint Cerf and Bob Kahn develop the basic communications protocols for the Internet.

1975
Bill Gates and Paul Allen form a partnership called Microsoft.

1976
The first Apple computer is produced, hand-built by Steve Wozniak. Queen Elizabeth sends an email, the first head of state to do so.

1980
Sinclair ZX-80 released for under £100.

1995
Amazon.com goes online and Ebay launches as 'AuctionWeb'. Sony release the PlayStation.

1996
Google launched (then known as 'BackRub') on Stanford university servers by Larry Page and Sergey Brin.

1996
IBM's Deep Blue computer defeats world chess champion Garry Kasparov.

1997
The NASA broadcast of images taken by the Pathfinder probe on Mars generates a record-breaking 46 million hits in one day.

2001
1M
Number of Internet hosts exceeds one million.

2001
Wikipedia debuts online.

2003
Skype released.

2004
Facebook founded by Mark Zuckerberg and a number of fellow Harvard students.

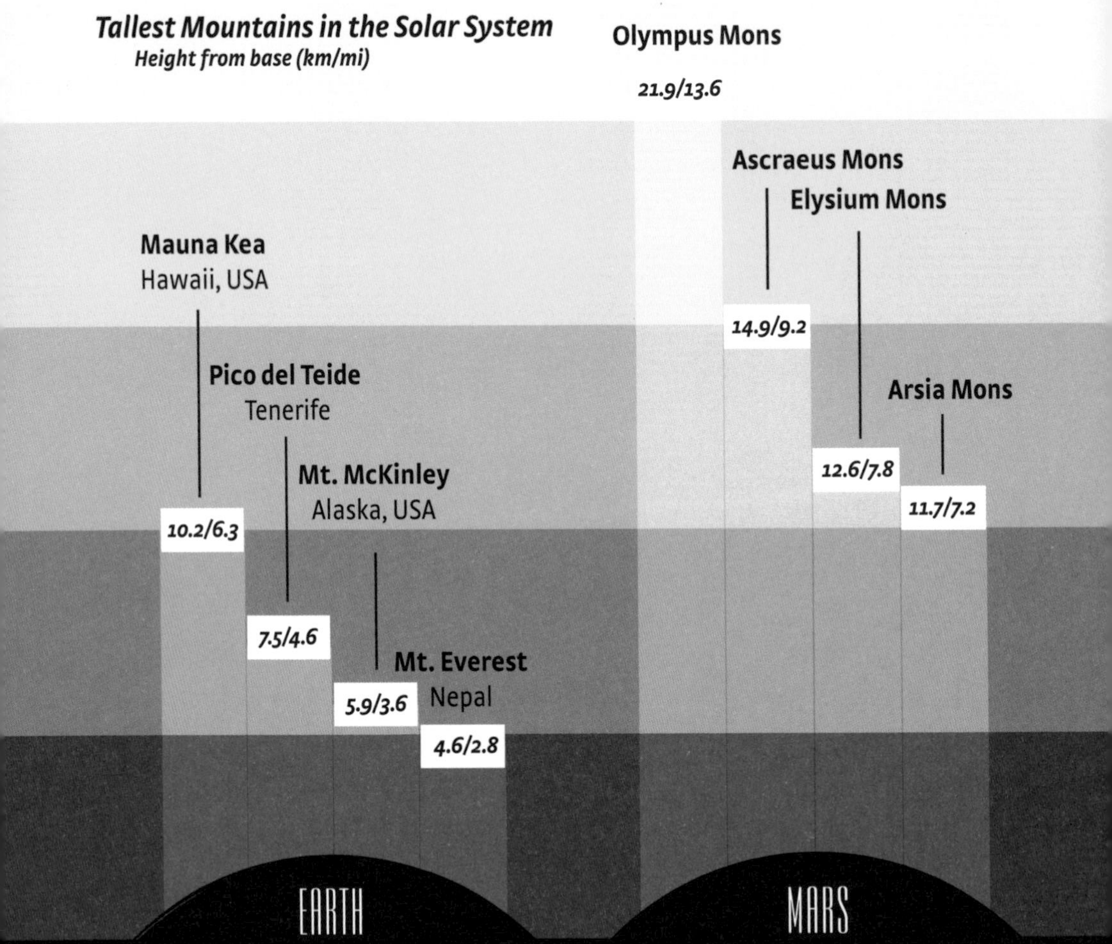

CLIMB EVERY **MOUNTAIN**

Mount Everest is the highest mountain in the world, when measured from sea level. But what if you take away the sea – how does it compare with mountains elsewhere on this planet and in the Solar System? If measured from base to summit, Everest only ranks as the fourth tallest mountain on Earth and looks pretty puny compared with the mountains of planet Mars, the asteroid Vesta, Jupiter's moon Io and Iapetus, one of Saturn's moons.

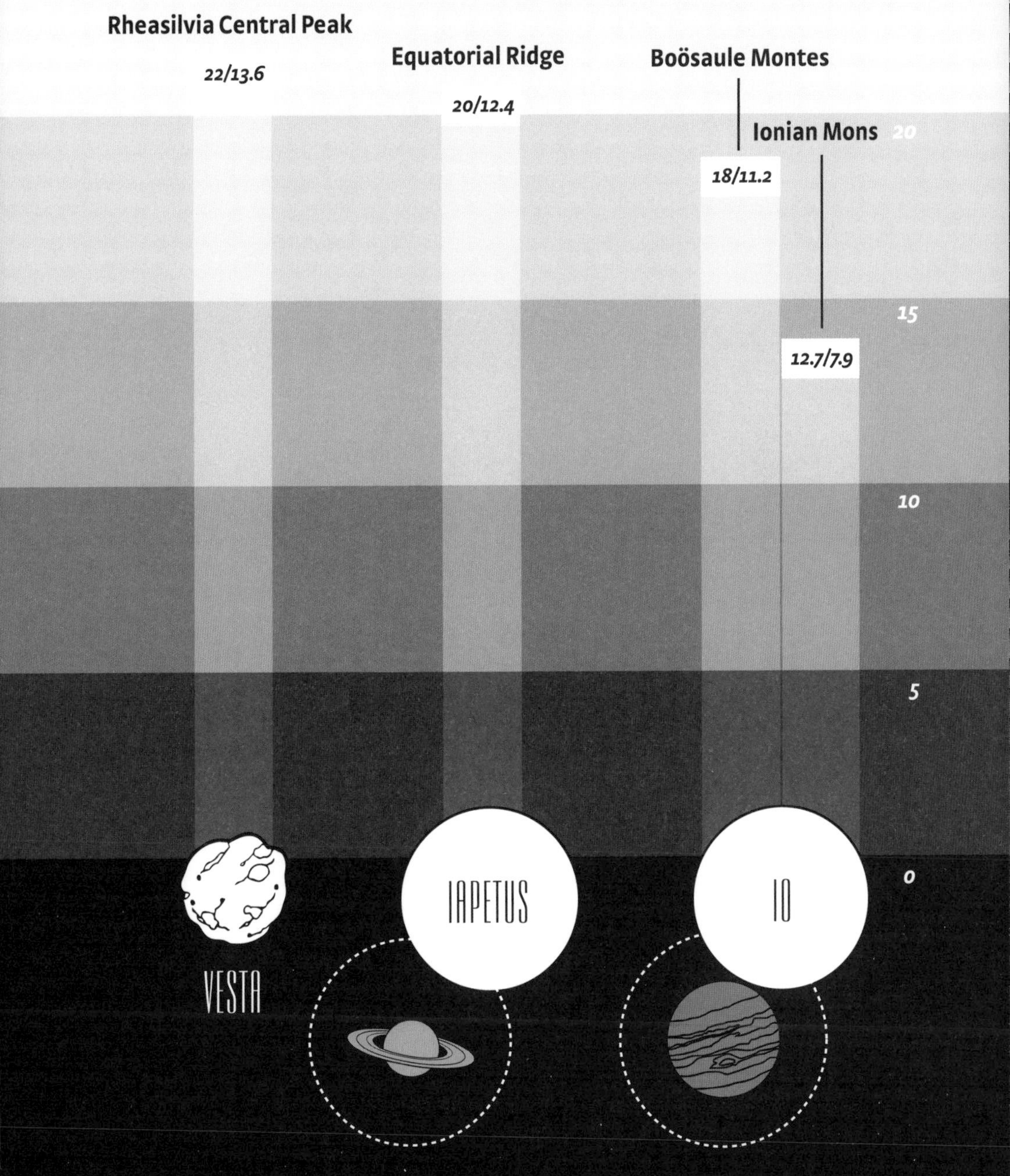
Rheasilvia Central Peak
22/13.6
Equatorial Ridge
20/12.4
Boösaule Montes
18/11.2
Ionian Mons
12.7/7.9
20
15
10
5
0
VESTA
IAPETUS
IO

THE WEIGHT OF **HAPPINESS**

Is it a myth that fat people are happy? By comparing the 2013 World Happiness Report (for the UN) ranking of countries with the World Health Organization's statistics on a country's body mass index average, we can get a good idea of whether that's true. The results suggest that there is an optimum BMI for happiness that is neither too fat nor too thin.

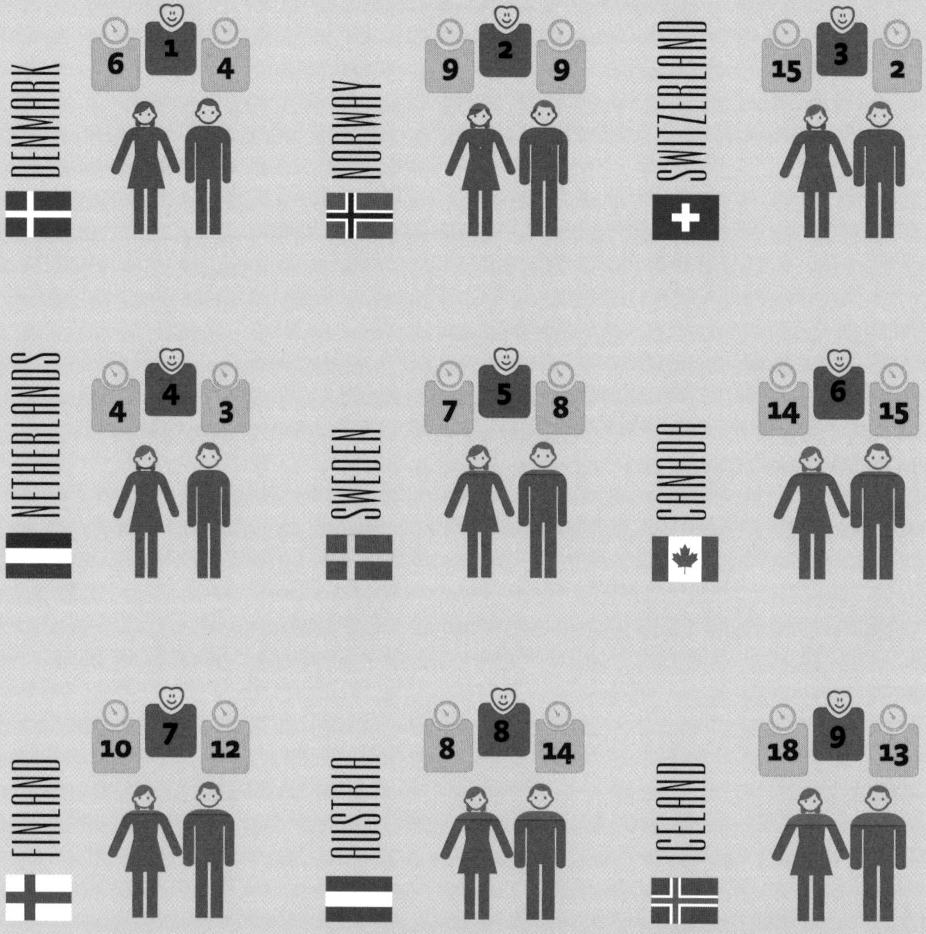

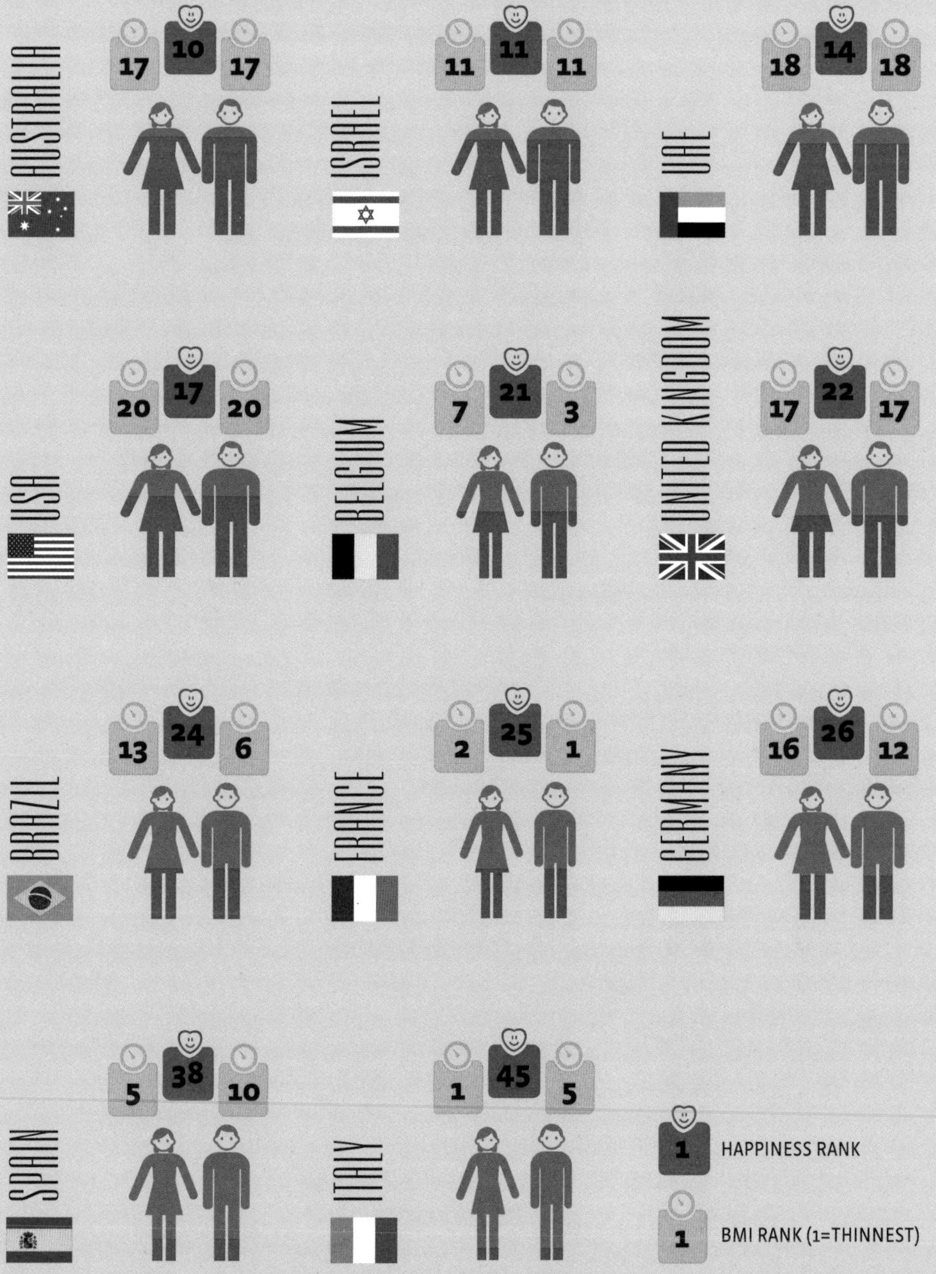
AUSTRALIA
17
10
17
ISRAEL
11
11
11
UAE
18
14
18
USA
20
17
20
BELGIUM
7
21
3
UNITED KINGDOM
17
22
17
BRAZIL
13
24
6
FRANCE
2
25
1
GERMANY
16
26
12
SPAIN
5
38
10
ITALY
1
45
5
1
HAPPINESS RANK
1
BMI RANK (1=THINNEST)

GREAT BALLS **OF FIRE**

Next time you're on a long-haul flight, think how lucky you are that it only takes 24 hours to traverse our globe. These celestial objects make Earth look like a dust mote in the universe.

THE EARTH

12,756 km
7,926 miles

The fourth smallest planet in the solar system.

JUPITER

Largest planet in Solar System

139,822 km
86,881 miles

x 11 = x 1

11 times larger than Earth in diameter, 318 times in mass.

THE SUN

Largest object in the Solar System

1.3 million km
808,000 miles

x 10 = x 1

10 times wider than Jupiter, 109 times wider than Earth.

UY SCUTI

Largest star discovered in the Scutum constellation

2.4 billion km
1.5 billion miles

x 1700 = x 1

More than 1700 times wider than the Sun.

SOLAR SYSTEM

In outer Solar System comets orbit at 1.87 light years from the Sun

4 light years

1 light year { 9.5 trillion km, 5.9 trillion miles

So large that a different scale is used.

MILKY WAY

A spiral galaxy consisting of 300 billion stars

100,000
light years wide

x 27,000 = x 1

27,000 times larger than Solar System.

MARKARIAN 348

(NGC262) the largest spiral galaxy known in Andromeda constellation

1.3 million
light years wide

x 13 = x 1

13 times as wide as Milky Way.

IC 1101

Supergiant elliptical galaxy in the Virgo constellation

6 million
light years wide

x 4.6 = x 1

nearly 5 times wider than Markarian 348, contains 100 trillion stars.

beforeitsnews.com, news.cnet.com, universetoday.com, wikipedia.org

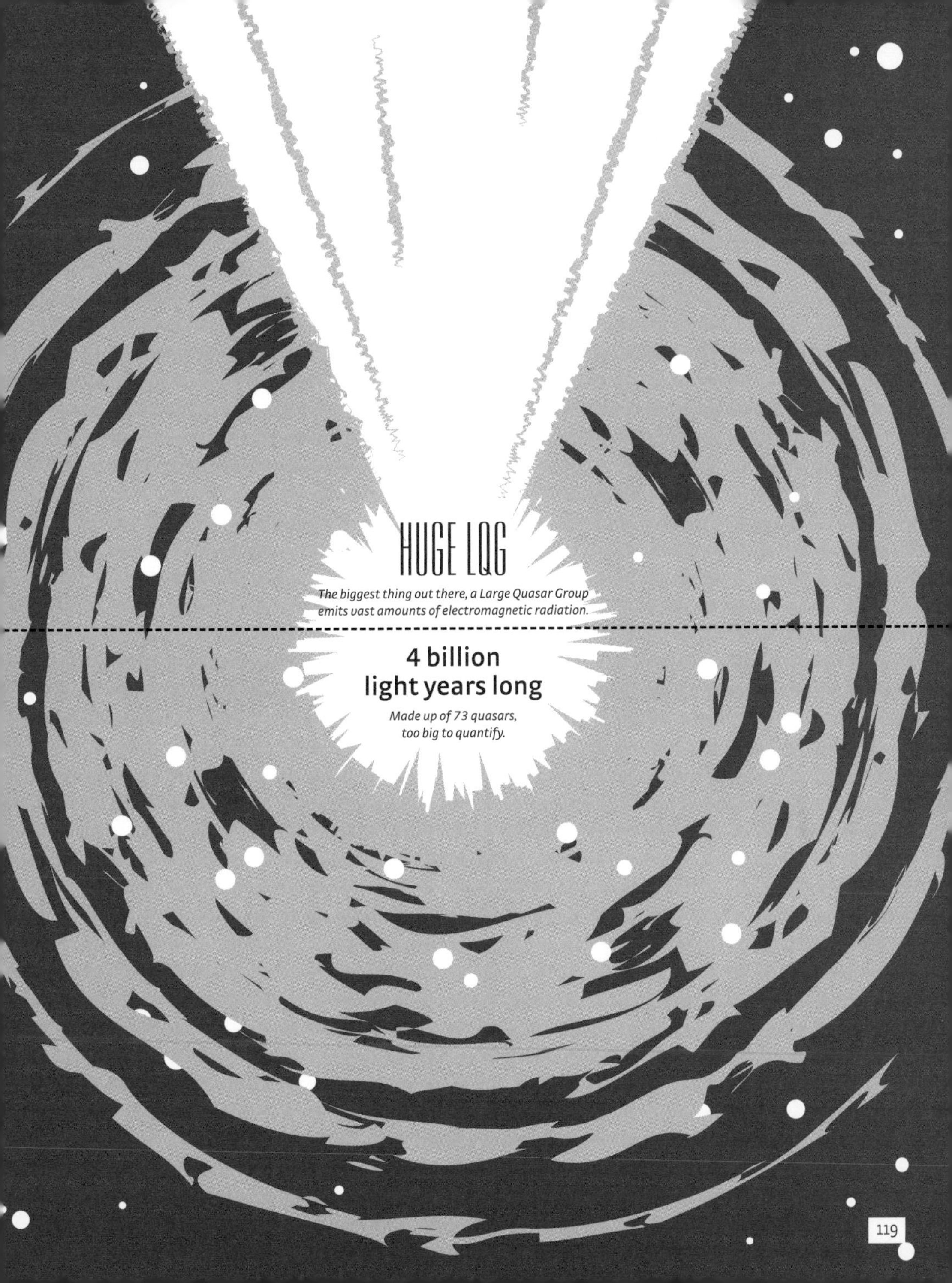
HUGE LQG
The biggest thing out there, a Large Quasar Group emits vast amounts of electromagnetic radiation.
4 billion light years long
Made up of 73 quasars, too big to quantify.

WHO REALLY **RUNS THE WORLD**

According to conspiracy theorists the world is 'run' by organizations that are not always secret, although their aims and actions are (of course). Here are the ten most powerful organizations in the world according to popular conspiracy theory, and the allegations of what they've done, or are attempting to do, to the world.

PRE - CE | **12TH CENTURY** | **1540** | **1717** | **1776**

REPTILIAN ALIENS

PRE-CHRISTIAN ERA

Covert control of humanity.

Senior US politicians, European royalty, show business stars

THE JESUIT BROTHERHOOD

CATHOLIC CHURCH

Pope as ruler of the world, end of democracy.

Former US Presidents, secretaries of state and major corporation bosses

THE ILLUMINATI

BAVARIA, GERMANY

Destruction of Christianity, maintenance of occultist beliefs, world rule.

Top bankers, US Presidents, CIA bosses, international politicians, British royal family

THE BLACK NOBILITY

VENICE, ITALY

Rule the world after de-populating it via wars.

European royal families, US presidential and oil-wealthy families

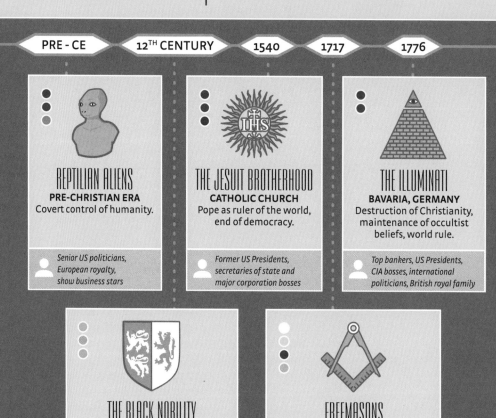

FREEMASONS

EUROPE

Lead a world government.

Royalty, astronauts, UK PMs, business leaders

ALLEGED ACTIONS

- Human abduction, sacrifice / ceremonies
- Satanic worship
- World leaders meet to indulge in pagan rites
- Began slave trade
- Child abuse
- Power behind rise of UK PMs / US Presidents
- Assassination: JFK, Pope John Paul II, Aldo Moro
- US High school shootings
- Caused 9/11 & other 'Islamist' attacks
- Controls US & UK governments, UN, Hollywood
- Mind control
- Runs the US Federal / World Bank & Trade Organizations
- Wars/revolutions, drugs trafficking & arms trades
- Invention and destruction of first atom bomb
- Responsible for AIDs epidemic
- Created Economic Crisis
- Formation of Illuminati
- Formation of the European Union
- Destabilize liberal and anti-capitalist governments
- Faked moon landings

1872 — 1897 — 1919 — 1948 — 1954

THE BOHEMIAN CLUB

CALIFORNIA, USA

Destruction of civilization, world population enslavement.

Unknown...

COUNCIL ON FOREIGN RELATIONS

USA

The establishment of free-market capitalism as the ruling world order.

Former US and UK senior politicians, business leaders, members of the Bilderberg Group

THE BILDERBERG GROUP

HOLLAND

Control of single world government, bank and army.

150 members

COMMITTEE OF 300 (THE OLYMPIANS)

EAST INDIA TRADING COMPANY

Ruling over a New World Order.

US politicians, major business leaders

CENTRAL INTELLIGENCE AGENCY OF THE USA

Protection and enhancement of **USA** via intelligence gathering.

21.5K employees

pseudoreality.org, armageddonconspiracy.co.uk, realworldorder.net, davidicke.com, wikipedia.org

IT'LL NEVER **CATCH ON**

Sometimes an idea is too advanced for the rest of the world to understand how important it might be. Tracing the time between an invention being made and its acceptance by society, we find that parachutes took three centuries, sliced bread took almost two decades, but Google was incorporated only nine years after the World Wide Web was first demonstrated.

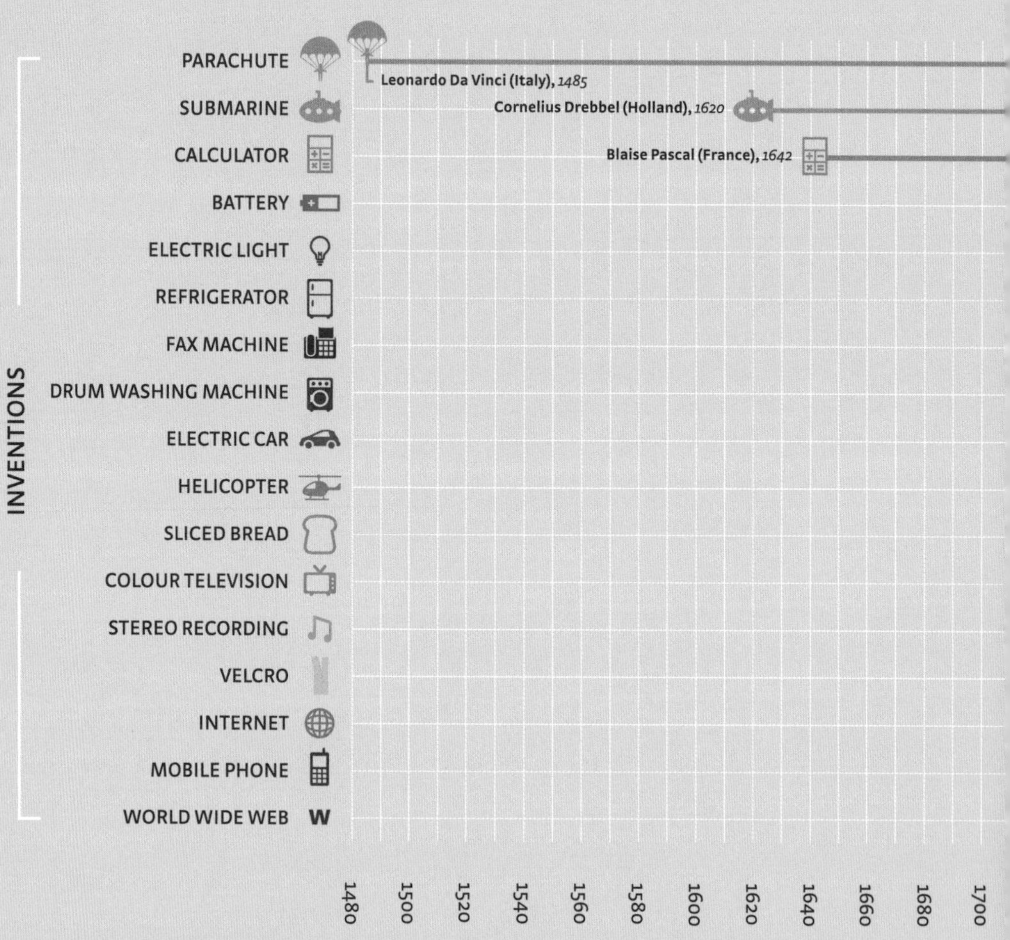

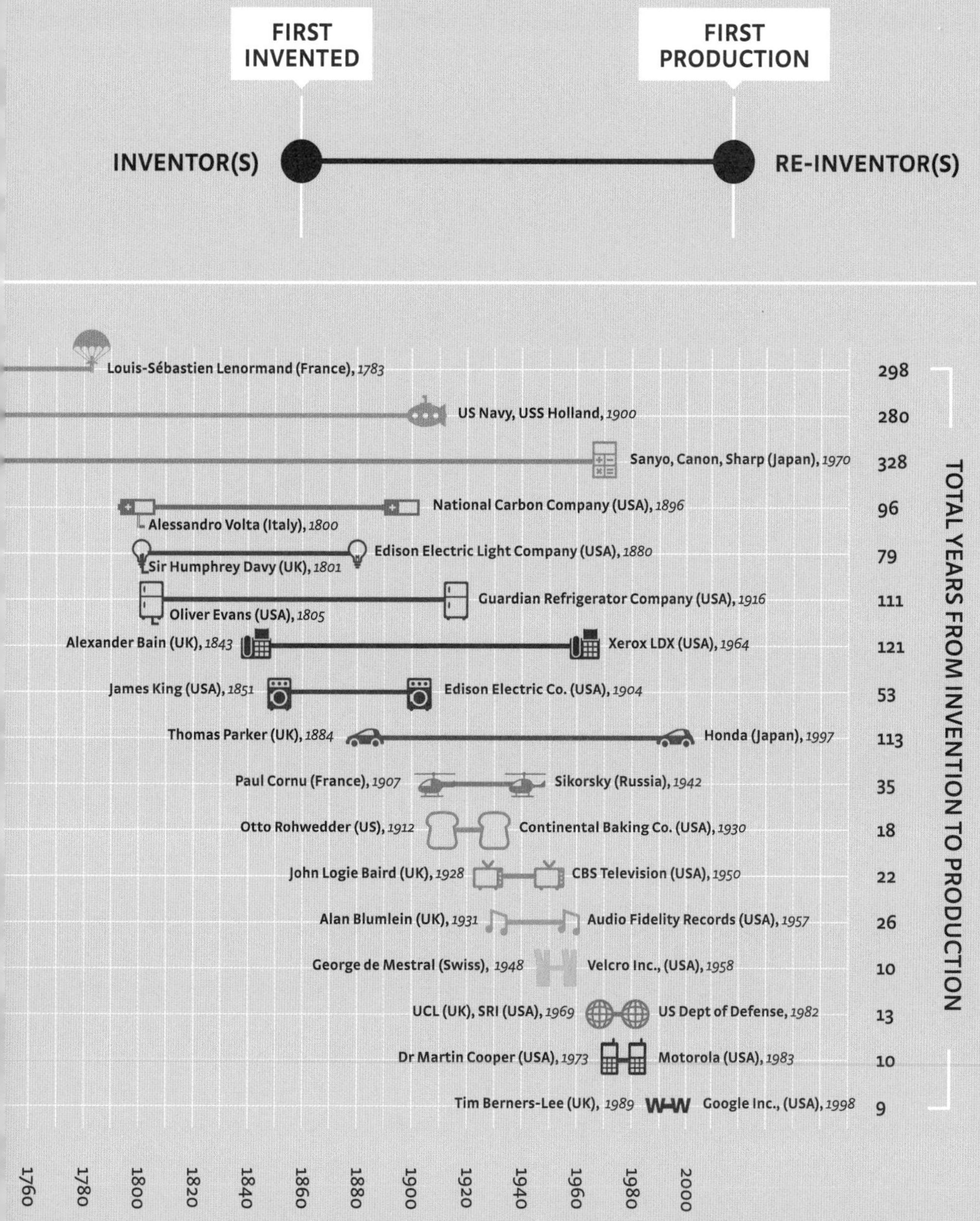

britishlibrary.co.uk, audiokarma.org, lightbulb.co.uk, wikipedia.org, Lemelson Foundation, American Physical Society, radiomuseum.org, Helicopters magazine, about.com, biography.com, velcro.co.uk, Gale Directory of Company Histories, Vintage Calculators Web Museum, ideafinder.com, bairdtelevision.com

ALT, SHIFT...**DELETE?**

The world's stores of data are kept underground or in vast storage units powered by electricity in often secret locations, and are dependent on energy resources never running out. There are still some old-fashioned libraries which hold acres of data in paper form, too. Here's how the books stack up against the computers and the threat against each's existence.

UK
THE BRITISH LIBRARY
170M items
112,000 m²

USA
LIBRARY OF CONGRESS
151.8M items
186,000 m²

USA
NEW YORK PUBLIC LIBRARY
53.1M items
60,079 m²

Russia
RUSSIAN STATE LIBRARY
44.4M items
57,600 m²

Japan
NATIONAL DIET LIBRARY
35.6M items
74,900 m²

China
NATIONAL LIBRARY OF CHINA
31.2M items
80,000 m²

PRIMARY THREATS TO LIBRARIES

- fire
- bombing
- earthquakes
- budget cuts
- tsunamis
- sacking
- lack of maintenance
- change in air conditions
- societies keen on book burning

GOOGLE
Goose Creek Berkeley County SC, USA
(Unknown)

GOOGLE
Lenoir NC, USA
(Unknown)

GOOGLE
Council Bluffs IA, USA
10,684 m² (35,052 ft²)

GOOGLE
Douglas County GA, USA
(Unknown)

GOOGLE
Mayes County OK, USA
130,064 m² (426,719 ft²)

GOOGLE
The Dalles OR, USA
15,236 m² (49,986 ft²)

GOOGLE
Changhua County, Taiwan
(Unknown)

GOOGLE
Singapore
(Unknown)

GOOGLE
Hamina Finland
(Unknown)

GOOGLE
St Ghislain Belgium
(Unknown)

GOOGLE
Dublin Ireland
(Unknown)

GOOGLE
Quilicura Chile
(Unknown)

IBM OPERATES 7.4 M² (8M FT²) OF DATA CENTERS ON 6 CONTINENTS

OTHER POSSIBLE GOOGLE DATA CENTERS

Google Mountain View CA, USA **(Unknown)**	*Google Pleasanton CA, USA* **(Unknown)**	*Google San Jose CA, USA* **(Unknown)**	*Google Los Angeles CA, USA* **(Unknown)**	*Google Palo Alto CA, USA* **(Unknown)**	*Google Portland OR, USA* **(Unknown)**	*Google Atlanta GA, USA* **(Unknown)**
Google Houston TX, USA **(Unknown)**	*Google Toronto Canada* **(Unknown)**	*Google Berlin Germany* **(Unknown)**	*Google Frankfurt Germany* **(Unknown)**	*Google Munich Germany* **(Unknown)**	*Google Zurich Switzerland* **(Unknown)**	*Google Groningen Netherlands* **(Unknown)**

LIBRARIES

WORLD'S LARGEST DATA CENTERS

GOOGLE DATA CENTERS which they admit to owning

FACEBOOK DATA CENTERS

Range Int'l Information Hub
Langfang, China
585,289 m² (6.3M ft²)

Switch SuperNAP
Las Vegas NV, USA
204,870 m² (2.2M ft²)

Data Center NSA
Bluffdale UT, USA
92,903 m² (1M ft²)

350 E Cernak
Chicago IL, USA
104,300 m² (1.1M ft²)

QTS Metro Data CTR, Atlanta GA, USA
91,974 m² (990,000 ft²)
(inc. Twitter)

Tulip Data City
Bangalore India
83,613 m² (900,000 ft²)

NAP Of The Americas
Miami FL, USA
69,677 m² (750,000 ft²)

Next Generation Data
Newport, Wales
69,677 m² (750,000 ft²)

Phoenix One
Phoenix AZ, USA
66,700 m² (718,000 ft²)

Microsoft
Chicago IL, USA
65,032m² (700,000 ft²)

Microsoft
Dublin, Ireland
54,255 m² (550,000 ft²)

DuPont Fabros
Elk Grove Village IL, USA
45,058 m² (485,000 ft²)

Microsoft
Quincy WA, USA
43,664 m² (470,000 ft²)

Microsoft
San Antonio TX, USA
43,664 m² (470,000 ft²)

FACEBOOK
Altoona IA, USA
44,222 m² (476,000 ft²)

FACEBOOK
Prineville OR, USA
28,521 m² (307,000 ft²)

FACEBOOK
Forest City NC, USA
27,871m² (300,000 ft²)

FACEBOOK
Lulea Sweden
27,000 m² (290,000 ft²)

PRIMARY THREATS TO DATA CENTERS

google.com, datacenterknowledge.com, equipemicrofix.com, govtech.com, newsroom.fb.con, wikipedia.org

Google
Reston VA, USA
(Unknown)

Google
Virginia Beach VA, USA
(Unknown)

Google
Tokyo, Japan
(Unknown)

Google
Seattle WA, USA
(Unknown)

Google
Chicago IL, USA
(Unknown)

Google
Miami FL, USA
(Unknown)

Google
Ashburn VA, USA
(Unknown)

Google
Mons Belgium
(Unknown)

Google
Eemshaven Netherlands
(Unknown)

Google
Paris France
(Unknown)

Google
London UK
(Unknown)

Google
Milan Italy
(Unknown)

Google
Moscow Russia
(Unknown)

Google
Sao Paolo Brazil
(Unknown)

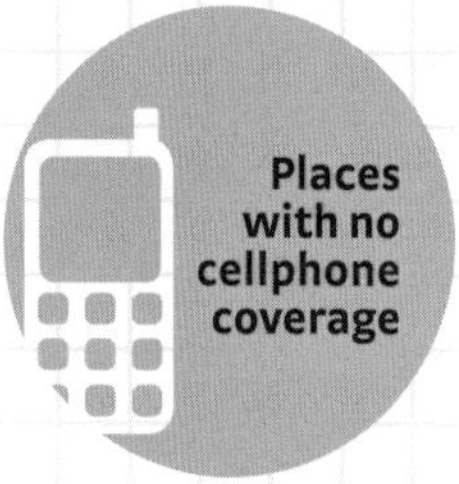

GOING OFF THE **GRID**

If you really want to escape the world of constant electronic attention, easy access to civilization or civilization in general, there aren't many remote places left in the world. Here are the world's ten least accessible, ten least populated and the ten places with no cellphone coverage and little electricity.

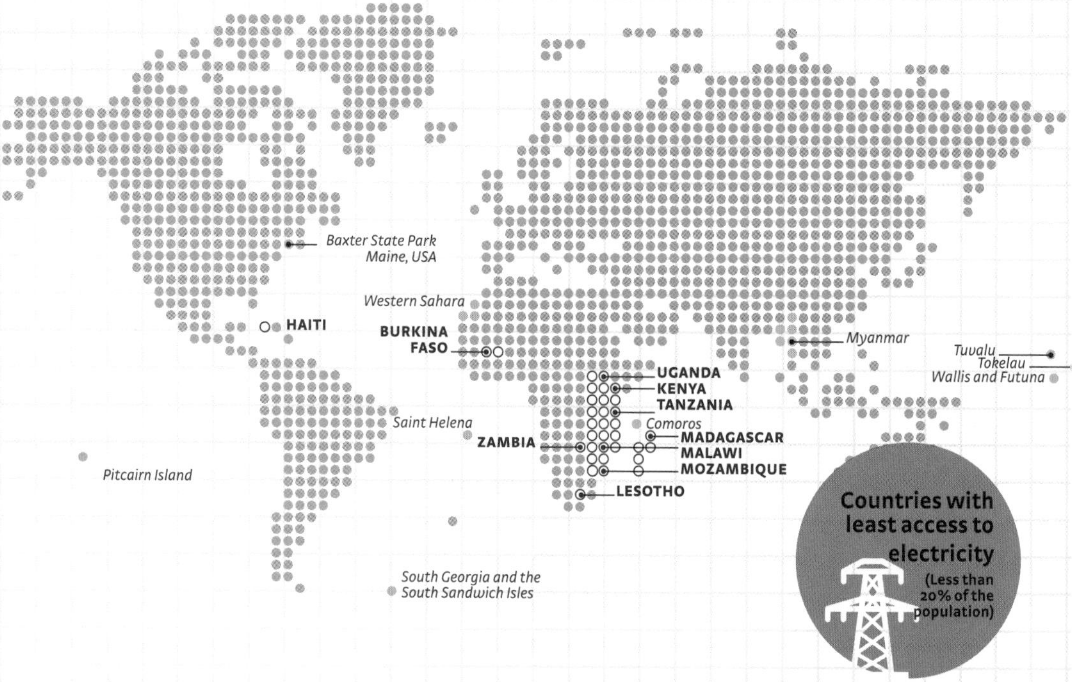

COUNTRY /LOCATION	POPULATION
South Georgia and the South Sandwich Isles	60
Pitcairn Islands, South Pacific	67
Tokelau, South Pacific	1,400
Saint Helena, South Atlantic	4,255
Tuvalu, Pacific Ocean	10,837
Wallis and Futuna, South Pacific	13,484
Baxter State Park, Maine, USA	63,000
Western Sahara, North Africa	513,000
Comoros, Indian Ocean	798,000
Myanmar, South East Asia	61,120,000

COUNTRY	% POP WITH ELECTRICITY	POPULATION
Tanzania, Africa	14.8	44,928,923
Kenya, Africa	18.1	44,037,656
Uganda, Africa	8.5	35,873,253
Mozambique, Africa	15	23,929,708
Madagascar, Indian Ocean	17.4	22,005,222
Malawi, Africa	8.7	16,407,000
Burkina Faso, West Africa	14.6	15,730,977
Zambia, East Africa	18.5	14,309,466
Haiti, Caribbean	20	9,719,932
Lesotho, South Africa	17	2,067,000

Most remote places
(No airport or roads)

COUNTRY /LOCATION	POPULATION
Pyramid Island, Kenya (in Lake Victoria)	0
Macquarie Island, South West Pacific Ocean	20–40
Kerguelen Islands, Southern Indian Ocean	50–100 (scientists)
Pitcairn Island, South Pacific	67
Supai, Arizona, USA	208
Tristan da Cunha, South Atlantic Ocean	less than 300
Ittoqqortoormiit, Greenland	452
Motuo county Nyingchi Prefecture, Tibet	less than 10,000
Cape York Peninsula, Australia	less than 18,000

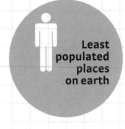

Least populated places on earth

COUNTRY	POPULATION	AREA (SQ.KM/SQ.M)	POP DENSITY (SQ.KM/SQ.M)
Canada	32,805,000	9,976,970/3,851,807.61	3/9
Australia	20,090,400	7,686,850/2,967,908.16	3/7
Libya, North Africa	5,765,600	1,759,540/679,361.91	3/8
Mongolia, Central Asia	2,791,300	1,556,000/604,249.63	2/5
Namibia, West Africa	2,030,700	825,418/318,695.54	2/6
Botswana, Africa	1,640,100	600,370/231,804.06	3/7
Mauritania, North Africa	3,086,900	1,030,700/397,955.33	3/8
Guyana, South America	765,300	214,970/83,000.35	4/9
Suriname, South America	438,100	163,270/63,038.87	3/7
Iceland	296,700	103,000/39,768.51	3/7

data.worldbank.org, worldtimezone.com, calibryze.com, wikipedia.org, worldatlas.com

TO INFINITY... **AND BEYOND?**

The next breakthrough in computer science will involve machines that can make quantum calculations. One already exists, claims a company who are selling such a device, and everyone's calling it the Infinity Machine. They cost $10 million each. Here's what we know about it.

THE THEORY

Quantum superposition calculates different outcomes simultaneously that can be opposite and contradictory: a thing neither is nor isn't, but is and isn't at the same time.

Schrödinger's cat

Schrödinger's cat is sealed in a box with a vial of poison and an object of radiation. If the radiation breaks the vial the cat will die. If it doesn't the cat will live. Schrödinger stated that the cat was both alive and dead in superposition until the box was opened.

THE COMPUTER

Traditional computers work in binary form, using units of either 1 or 0

D-WAVE 2 uses 512 tiny superconducting circuits known as 'qubits' to calculate using both 1 and 0 simultaneously

Several possible answers to questions are graded by optimization

Because it can perform several tasks at the same time, the D-WAVE 2 is infinitely faster than standard computer.

In a comparison test May 2013 a 439-qubit D-WAVE 2 computer performed 3600 times faster than the fastest traditional computer; found 100 + variables in 0.5 seconds

FUNDERS INCLUDE

Draper Fisher Jurvetson
venture capitalists who invested in Skype and Tesla cars

Jeff Bezos
founder of Amazon.com
Q-Tel, investment arm of the CIA

CUSTOMERS

Lockheed Martin
air and defense contractors
NASA
Google
Universities Space Research Association

FORM — 5 BUILT

A 3m black box with cylindrical cooling tower inside.

Operates at -273°C, which is colder than outer space (-272°C) and 0.02° above absolute zero.

It is shielded to 50,000 times less than Earth's magnetic field

It is in a high vacuum where pressure is 10 billion times lower than atmospheric pressure

Includes 192 in, out, and control lines from room temperature to the chip

'The Fridge' cooling system and servers consume 15.5kW of power (traditional super computers use 2,335kW)

MAKER: D-WAVE, founded 1999, based in Vancouver BC, Canada, Washington DC and Palo Alto, CA, USA

LIFE, DEATH AND **WEALTH**

What relation does a country's wealth have on the life expectancy and mortality rate of its people? Here are the GDP per capita measurements for the ten most significant countries of the world, set against their child and adult mortality rates plus life expectancy.

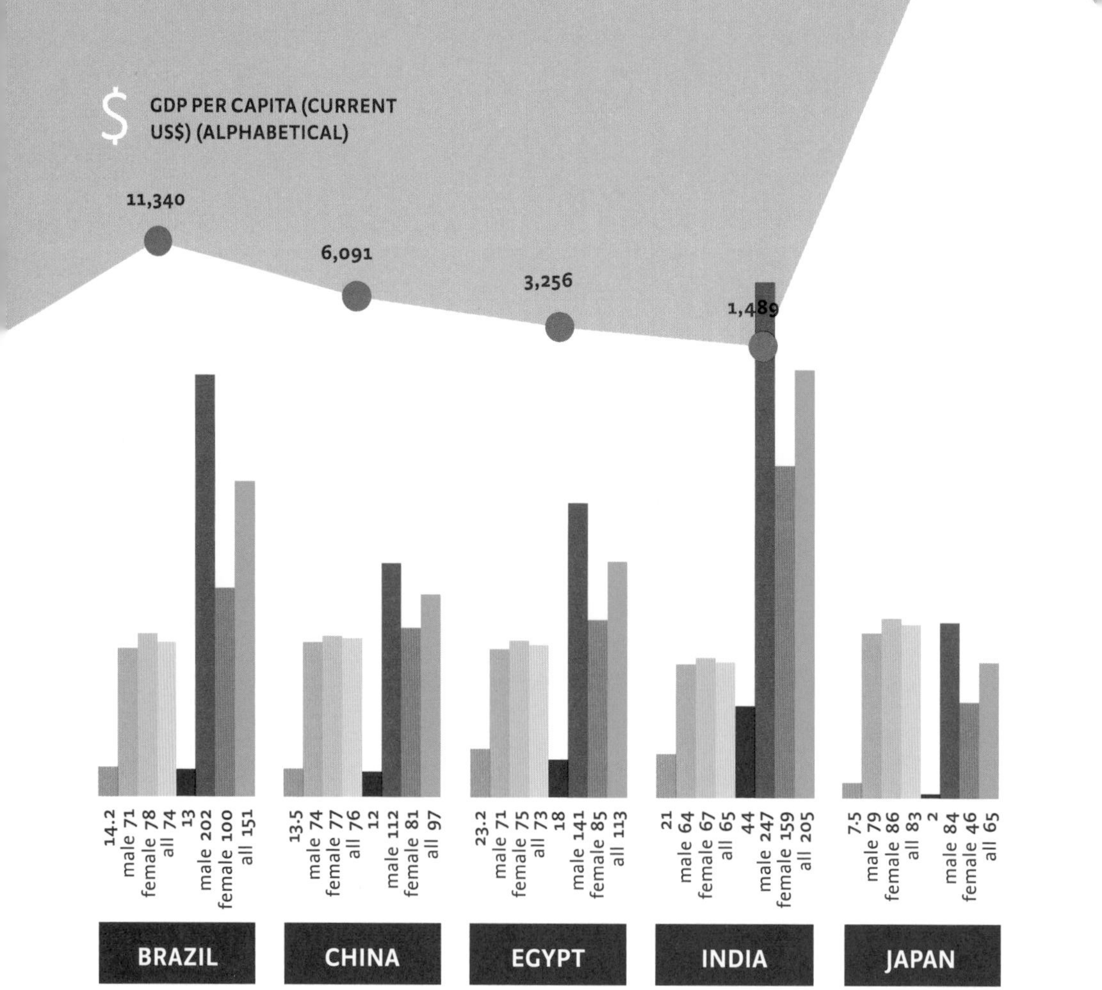

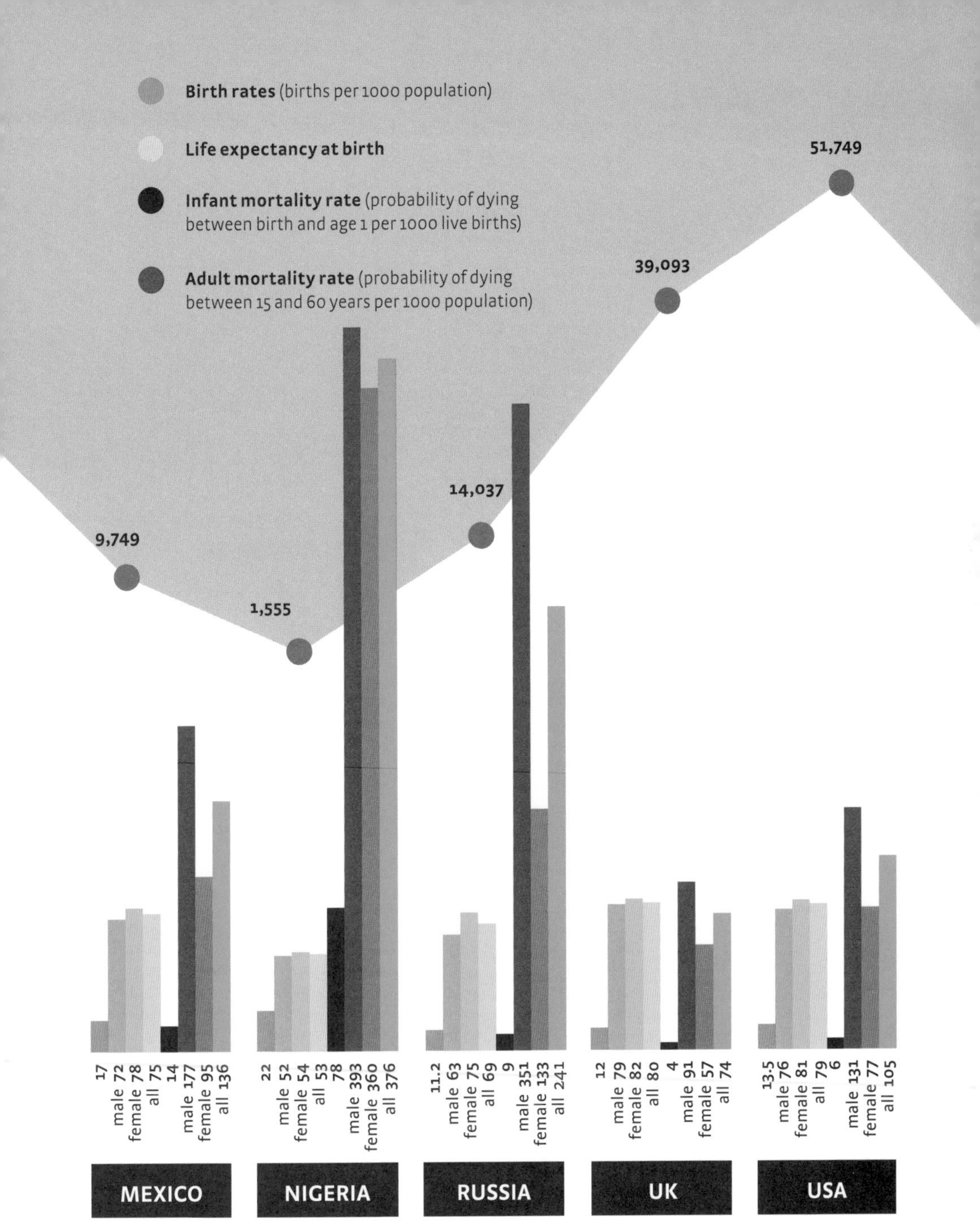

Gapminder/UN Population Division, World Health Organization, data.worldbank.org

INSIDE THE **LARGE HADRON COLLIDER**

Scientists are convinced that the Large Hadron Collider will provide them with details of what went on at the moment of the Earth's creation. But what exactly is the LHC? Here are the major facts.

CIRCUMFERENCE: 27km (16.8 miles)

LARGE: the biggest and most powerful particle accelerator in the world.

HADRON: accelerates protons or ions – types of hadron.

COLLIDER: particles make up two beams that travel in opposite directions and collide at four points.

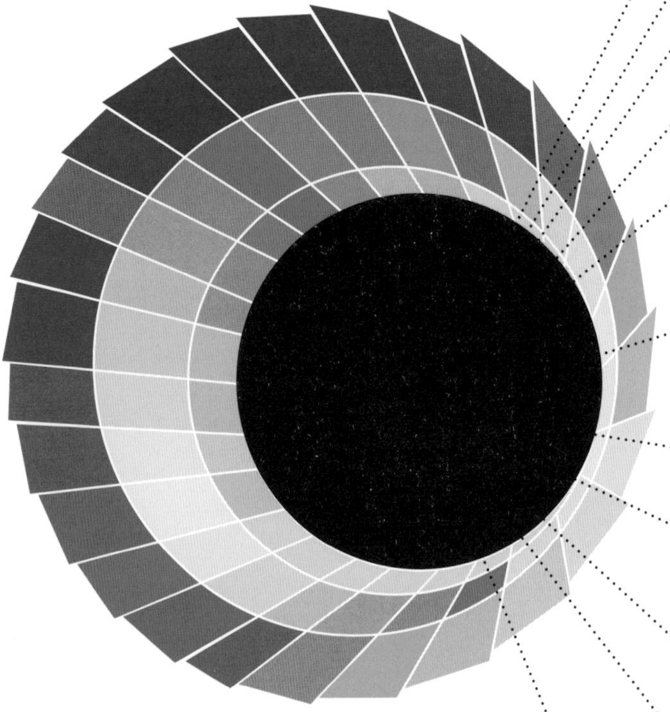

- Max speed: up to 0.999999991 times the speed of light, more than 11,000 times per second.
- Collisions per second: 600 million (est.)
- At full energy the LHC's proton beams are as powerful as a train at 150km/h (93mph)
- Operating temperature: -271.3°C (around -270°C), colder than deep outer space.
- Uses enough 0.007 mm thick, superconducting niobium-titanium filaments, (10 times thinner than a human hair) to stretch to the Sun and back more than five times.
- Two nanograms of hydrogen are accelerated each day. It would take approximately one million years to accelerate 1 gram of hydrogen.
- Pressure in beam pipes is approx. ten times lower than on the Moon.
- The CMS (Compact Muon Solenoid), a detector, uses a huge solenoid magnet to bend the paths of particles from collisions in the LHC.
- The magnets of the CMS contain more iron than the Eiffel Tower – around 10,000 tonnes.
- Major experiments at the LHC produce enough data annually to fill around 100,000 DVDs.

DEPTH: 175m to 50m

SNARLING UP **TRAFFIC**

If you've ever driven on a major road for any distance you'll know that there are times when a jam occurs, and you slow to a crawl for a while and then pick up speed again. After a few miles of travelling at normal speed you don't see any crash or sign of what caused the earlier snarl-up. That's because this is (probably) what happened:

1. Two cars travelling at steady high speed of 70mph/110kh, one behind the other when the front brakes for no reason

   Second car slows to 60mph/95kh

2. cars 1/4 mile (400m) behind front cars slow to 45mph/70kh

3. cars behind the second wave (1/2m, 800m from 1st) slow to 30mph/30kh on seeing brake lights in the distance

4. cars in fourth wave (1.5m, 2.2k behind 1st) slow sharply at apparent build-up ahead, to 15mph/25kh

5. original cars in first row of traffic have resumed speed at 70mph/110kh

6. cars up to 2 miles (3.2k) behind the second wave are crawling (5mph/8k) in traffic build-up until front lines regain speed

HAPPY **ACCIDENTS**

Sometimes things don't go as planned, even for professionals. Sometimes accidents happen to scientists and doctors that prove to be far more fortuitous than if the original experiment begun had not gone 'wrong'. Here are some of the world's happiest accidents and accidental discoveries, all of which have changed the world we live in.

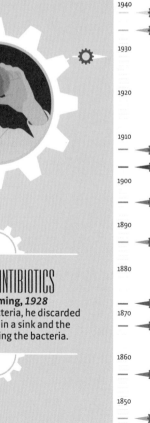

PENICILLIN ANTIBIOTICS

Sir Alex Fleming, *1928*
Whilst studying bacteria, he discarded a dish of bacteria in a sink and the mold settled, killing the bacteria.

Scotchgard Dirt Repellant: Pasty Sherman, *1970*
Whilst working on an artificial rubber, she spilled an experimental chemical on her shoe and noticed her shoe stayed clean.

Heart Pacemaker: Wilson Greatbatch, *1956*
Whilst walking with a heart monitor, he plugged the wrong transistor into the machine and heard a heartbeat sound.

Velcro Fastener: George du Mestral, *1955*
Whilst walking his dog, he wondered why seeds stuck so tightly to his dog's coat.

Super Glue Adhesive: Harry Coover, *1951*
Whilst working on a jet canopy, he added cyanacrylate to a plastic polymer but it was too sticky to use.

Slinky Toy: Richard Jones, *1943*
Whilst working on a power monitor, he dropped a spring and watched it move slowly around the room.

Teflon Non-Stick Coating: Roy Plunkett, *1938*
Whilst testing refrigerants, an experimental fridge gas made the inside of the gas canisters non-stick.

Cat's Eyes Reflectors: Percy Shaw, *1934*
Whilst trying to get home, he drove in fog by following his lights shining on tram rails.

Bakelite Plastic: Leo Baekeland, *1907*
Whilst working on insulation, he tried an experimental mixture which was malleable but dried hard.

Safety Glass: Édouard Bénédictus, *1903*
Whilst working with plastic resin, he dropped a resin-filled beaker and the broken shards of glass stuck together.

X-Ray Camera: Wilhelm Röntgen, *1895*
Whilst working on the cathode ray tube, he electrified a tube and saw a nearby screen glowing in the dark.

Vaseline Petroleum Jelly: Robert Chesebrough, *1872*
Whilst working with oil residue, he saw oil workers using the oily residue left on the drills to treat burns.

Dynamite Explosive: Alfred Nobel, *1867*
Whilst working with nitroglycerine, he observed that it leaked into packaging and made the explosive stable.

Purple Textile Dye: Sir William Perkin, *1856*
Whilst working on artificial quinine, he experimented with coal tar residues which produced a vivid mauve.

2000 1990 1980 1970 1960 1950 1940 1930 1920 1910 1900 1890 1880 1870 1860 1850

SMART DUST SENSORS

Jamie Link, ***2003***
Whilst working with semiconductors, a silicon chip shattered and she noticed individual pieces still worked.

POST-IT NOTES

Spencer Silver, ***1968***
Whilst working on a strong glue, he made a non-drying glue which a colleague used to keep a bookmark in place.

MICROWAVE OVEN

Percey Spencer, ***1945***
Whilst working on a radar device, he stood in front of the radar and felt his chocolate bar melting in his pocket.

COCA-COLA SODA DRINK

John Pemberton, ***1886***
Whilst working with painkillers, he accidentally added soda water to a herbal headache cure.

MEDICAL ANAESTHESIA

Dr Horace Wells, ***1846***
Whilst working as a dentist, he witnessed a stage show use of nitrous oxide (laughing gas).

businessinsider.com, geniusstuff.com, howstuffworks,com, rdasia.com, csmonitor.com, inc.com

GOING UP
AND DOWN

Where would Manhattan or Tokyo be without the elevator? Major high-rise living requires smooth-running, properly functioning elevators, and in the 21st century most are. But then, as this timeline demonstrates, it's taken more than 2000 years of development to get here.

Most angled elevator
Luxor Hotel, Las Vegas, Nevada, USA

39° angle

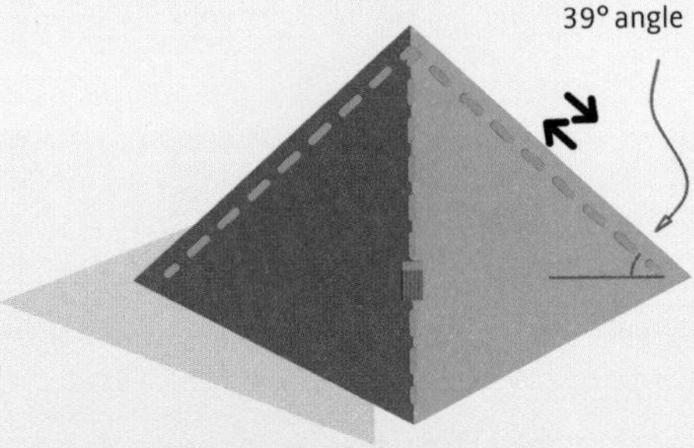

236BC Archimedes builds first elevator

Fastest + Highest elevators
Burj Khalifa, Dubai, UAE

10m/33ft per second
max height 800+m/ 2625ft

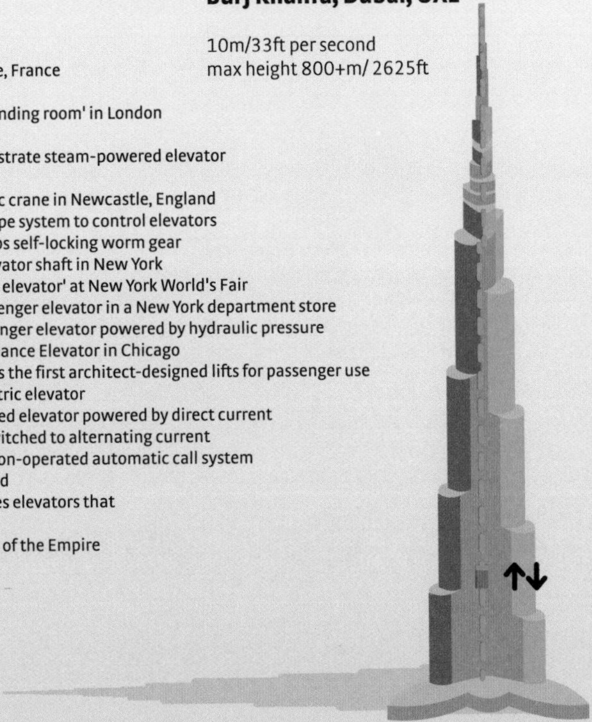

1743 'Flying chair' installed at Versailles Palace, France

1823 Architects Burton and Horner build 'ascending room' in London

1835 British inventors Frost and Strutt demonstrate steam-powered elevator

1845 Sir William Armstrong develops hydraulic crane in Newcastle, England
1850 Henry Waterman, New York, invents a rope system to control elevators
1851 George Fox & Co of Boston, Mass. develops self-locking worm gear
1853 First building designed to include an elevator shaft in New York
1854 Elisha Otis demonstrates the first 'safety elevator' at New York World's Fair
1857 Otis Elevator Company installs first passenger elevator in a New York department store
1867 Leon Edoux demonstrates the first passenger elevator powered by hydraulic pressure
1869 William E. Hale introduces the Water Balance Elevator in Chicago
1870 New York's tallest building (9 storeys) has the first architect-designed lifts for passenger use
1878 Siemens of Germany builds the first electric elevator
1889 Otis Elevator Company installs first geared elevator powered by direct current in a New York department store, then switched to alternating current
1924 Otis Elevator Company installs first button-operated automatic call system
1929 Automatic door opening system patented
1932 Empire State Building, New York, includes elevators that travel at 1000 feet per minute
1945 Longest elevator fall survived; 75 storeys of the Empire State Building by Betty Lou Oliver

1999 First pneumatic vacuum elevator

Tallest outdoor elevator
Bailong Elevator, Zhangjiajie, China

rises up against a steep cliff face 1000 feet high

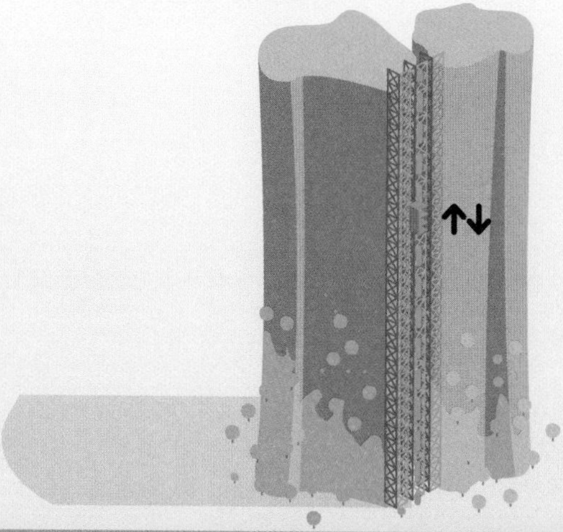

Largest elevator
Five elevators in Umeda Hankyu Building, Osaka, Japan

3.4m/ 11ft wide, 2.8m/ 9ft long, 2.6m/ 8.5ft high, 80 persons capacity

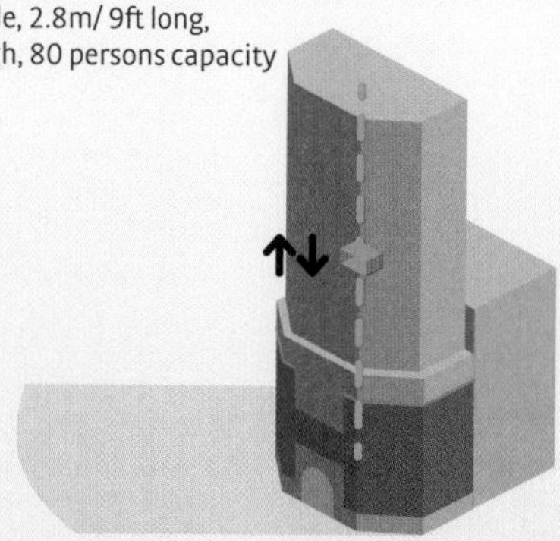

Deepest elevator
Mponeng Mine, Orange Free State, South Africa

3073m/ 9964ft (1200m/ 3937ft below sea level)

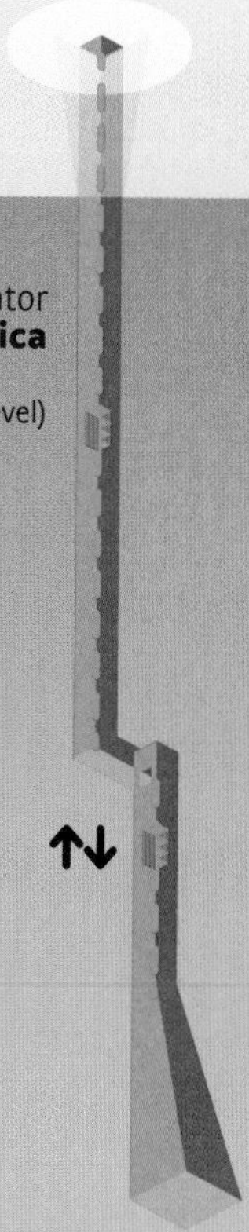

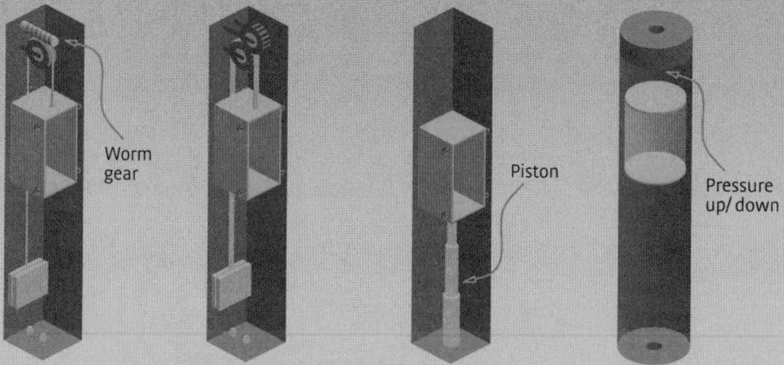

Traditional: AC or DC electric motors, worm gears to control a steel rope

Direct traction: AC or DC electric motors power gear wheels connected to toothed track inside elevator shaft

Hydraulic: Water or oil pressure to control piston rods

Pneumatic vacuum: Cable inside a tube, depressurizing the top of the tube sucks up the cabin, repressurization sends it down

theelevatormuseum.org, wikipedia.org

QUALIFIED TO RUN A COUNTRY?

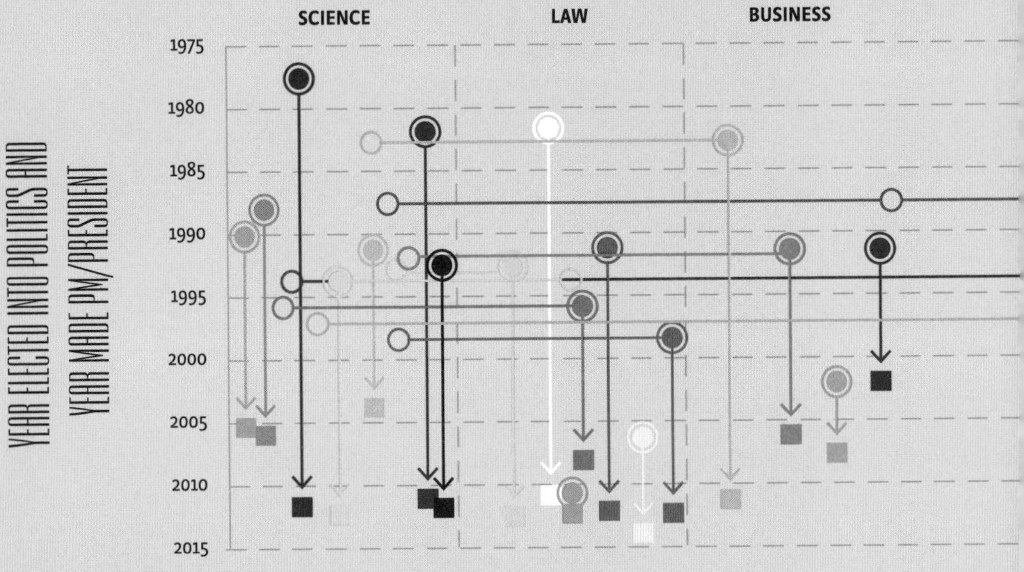

EMPLOYMENT PRE-POLITICS

AUSTRALIA: Tony Abbot born 1957	Catholic seminarian, journalist, business manager
AUSTRIA: Werner Faymann born 1960	Consultant at central bank, provincial chairman of Viennese Tenants Counselling
BRAZIL: Dilma Rousseff born 1947	Anti-capitalist guerilla, imprisoned 1970–73, political administrator
CANADA: Stephen Harper born 1959	MP aide
CHINA: Li Keqiang born 1955	Communist Youth League secretary, Peking University
DENMARK: Helle Thorning-Schmidt born 1966	None
FINLAND: Jyrki Katainen born 1971	Teacher
FRANCE: François Hollande born 1954	Presidential economic advisor
GEORGIA: Irakli Garibashvili born 1982	Director of a charity, board member of a bank, director of a record company
GERMANY: Angela Merkel born 1954	Chemistry research scientist
GREECE: Antonis Samaras born 1951	None
GREENLAND: Aleqa Hammond born 1965	Tourism guide, regional co-ordinator, commissioner Inuit Circumpolar Council
INDIA: Manmohan Singh born 1932	United nations administrator, governmental advisor
IRAQ: Nouri al-Maliki born 1950	Teacher, newspaper editor

What qualifications do you need in order to become the political ruler of a country? Tracing the educational qualifications and pre-political work experience of 30 elected leaders of major economic powers shows a distinct tendency toward certain educational subjects.

○ Degree/area of study

● Year elected into politics

■ Year made PM/President

DEGREE/AREA OF STUDY

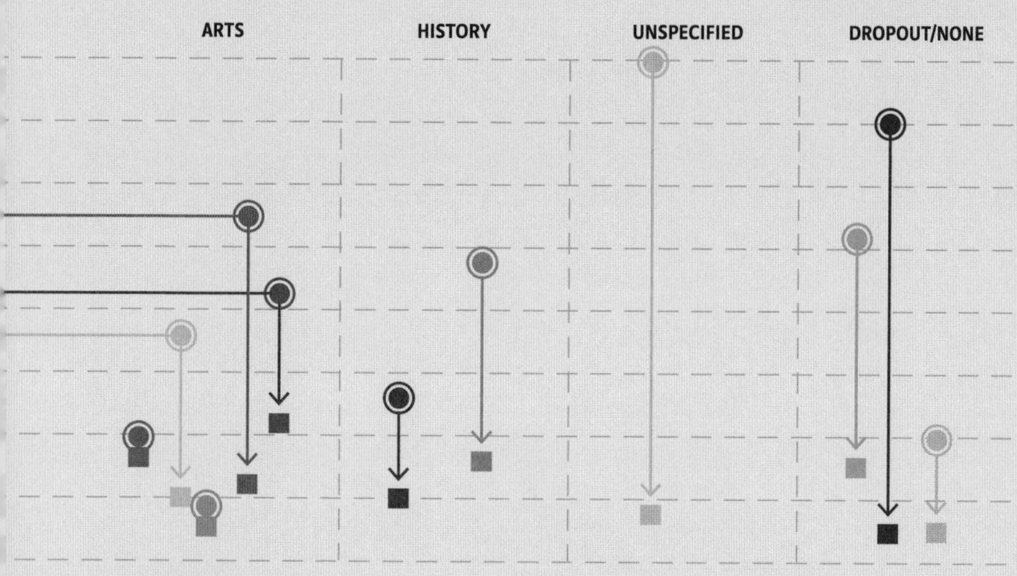

IRELAND: Enda Kenny born 1951	Primary school teacher
ISRAEL: Benjamin Netanyahu born 1949	Army captain, management consultant, director of NGO, marketing director, Israeli ambassador
ITALY: Matteo Renzi born 1975	Provincial secretary of Italian People's Party
JAPAN: Shinzo Abe born 1954	Executive at Kobe Steel, assistant to minister for Foreign Affairs, secretary to political party bosses
NORTH KOREA: Pak Pong-ju born 1939	Manager of food factory
SOUTH KOREA: Jung Hong-won born 1944	Legal prosecutor
LUXEMBOURG: Xavier Bettel born 1973	TV talkshow host, lawyer
NETHERLANDS: Mark Rutte born 1967	HR manager at Unlever, Calvé, director human resources for IgloMora Groep (Unilever)
NEW ZEALAND: John Key born 1961	Auditor, factory project manager, foreign exchange dealer
POLAND: Donald Tusk born 1957	Journalist
RUSSIA: Vladimir Putin born 1952	KGB officer, International Affairs officer, Leningrad State University, Mayor of Leningrad
SPAIN: Mariano Rajoy born 1955	Civil servant
SWEDEN: Fredrik Reinfeldt born 1965	Secretary at Stockholm Municipality Council
TURKEY: Recep Tayyip Erdogan born 1954	Semi-professional footballer, Mayor of Istanbul
UK: David Cameron born 1966	Director corporate afairs at a TV company, special advisor to Tory politicians
USA: Barack Obama born 1961	NY Public Interest Research group admin, director Chicago Developing Communities Project, president *Harvard Law Review*, senior lecturer in law University Chicago Law School, associate lawyer, legal counsel

WORLD OF **DINOSAURS**

When dinosaurs walked the Earth, where exactly did they amble? Well, we know where they perished, because our knowledge of different dinosaurs comes from archaeological digs at sites around the world. Here's where the twenty most prevalent and exciting dino skeletons have been found.

1 Tanzania	2 Portugal	3 Uruguay	4 Argentina	5 Madagascar
6 Germany	7 Namibia	8 Zimbabwe	9 NE China	10 Chile
11 S. India	12 Russia	13 N. Africa	14 S. Africa	15 Canada
16 United Kingdom	17 Belgium	18 Mongolia	19 Australia	20 Antarctic
21 USA				

● Carnivore ● Herbivore ⟼ Length Ɪ Height

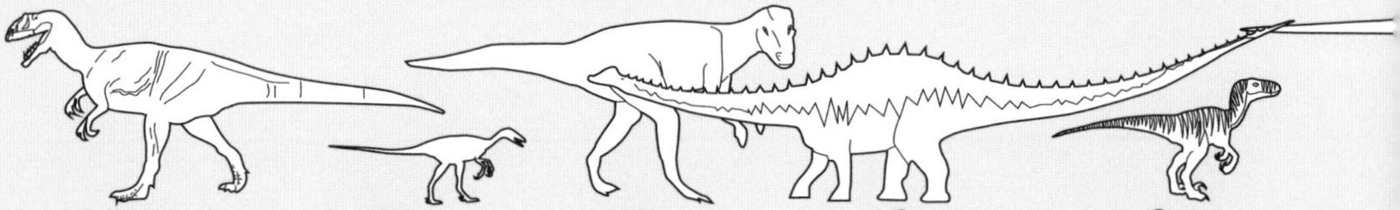

Megalosaurus (181–169 million years ago)
16
⟼ 9m (29.5ft)
Ɪ 3m (9.8ft)

Eoraptor (228)
4
⟼ 1m (3.3ft)
Ɪ 30cm (12in)

Hadrosaurus (80–79)
21 [1]
⟼ 10m (32.8ft)
Ɪ 3m (9.8ft)

Diplodocus (155–145)
21 [2]
⟼ 27m (88.5ft)
Ɪ 5m (16.4ft)

Velociraptor (85–80)
9 12 18
⟼ 2m (6.5ft)
Ɪ 1m (3.3ft)

Tyrannosaurus Rex (85–60)
15 18 21 [3]
⟼ 12m (39.4ft)
Ɪ 6m (19.6ft)

Stegosaurus (155–150)
5 21 [4]
⟼ 9m (29.5ft)
Ɪ 3m (9.8ft)

Majungasaurus (70–66)
5
⟼ 7m (23ft)
Ɪ 4m (13.1ft)

Muttaburrasaurus (113–97)
19 [5]
⟼ 7m (23ft)
Ɪ 3m (9.8ft)

Brachiosaurus (156–145)
1 2 21 [6]
⟼ 26m (85.3ft)
Ɪ 16m (52.5ft)

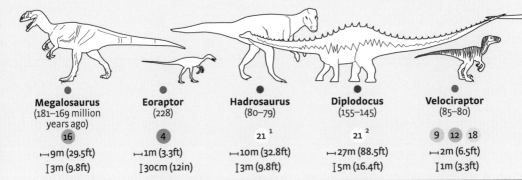

[1]New Jersey [2]Colorado, Montana, Utah, Wyoming [3]Montana, Texas, Utah, Wyoming, New Mexico [4]Colorado, Utah, Wyoming [5]Queensland [6]Colorado

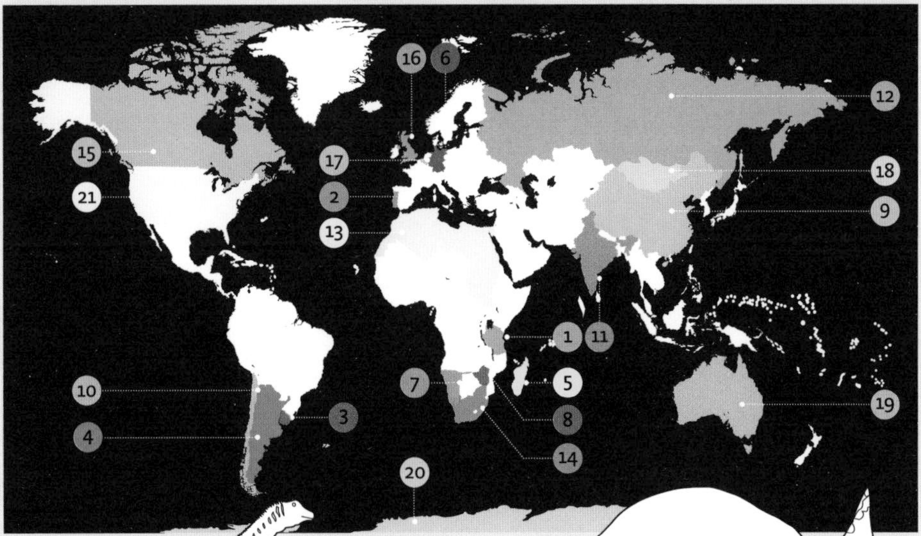

Antarctosaurus (80–65)	Sinosauropteryx (135–121)	Antarctopelta (74–70)	Iguanodon (135–125)	Triceratops (72–65)
3 4 10	9	20	6 13 16 21 17	15 21 (Western)
⟼19m (62.3ft)	⟼1.25m (4.1ft)	⟼4m (13.1ft)	⟼8m (26.2ft)	⟼2m (6.5ft)
I 7m (23ft)	I 29cm (11in)	I 2m (6.5ft)	I 2.7m (8.9ft)	I 9m (29.5ft)

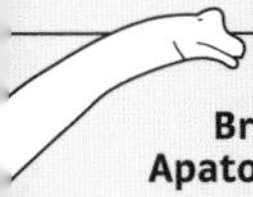

Brontosaurus/ Apatosaurus (157–146)	Pteranodon/ Pterodactyl (85–75)	Allosaurus (150–140)	Massospondylus (208–190)	Ankylosaurus (68–65.5)
21 [7]	16 21 [8]	1 2 19 21 [9]	8 7 14 21 [10]	15 21 [11]
⟼27m (88.5ft)	⟼10m (32.8ft)	⟼12m (39.3ft)	⟼1.8m (5.9ft)	⟼10.6m (35ft)
I 4.6m (15ft)	I 1.8m (5.9ft)	I 3m (9.8ft)	I 4m (13.1ft)	I 1.2m (4ft)

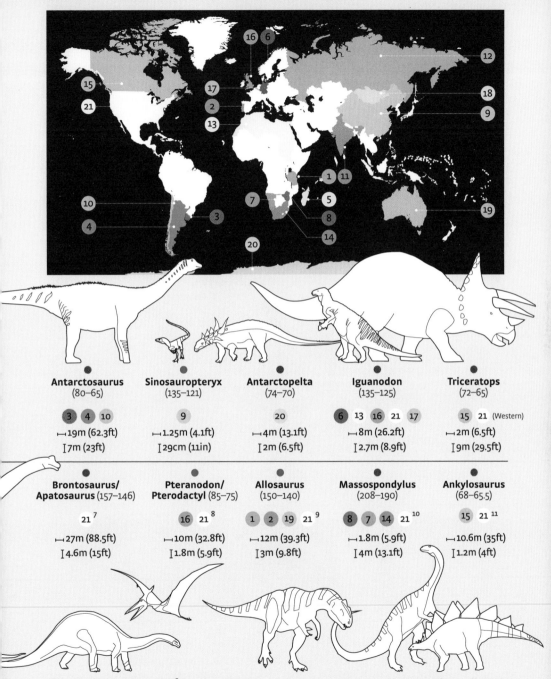

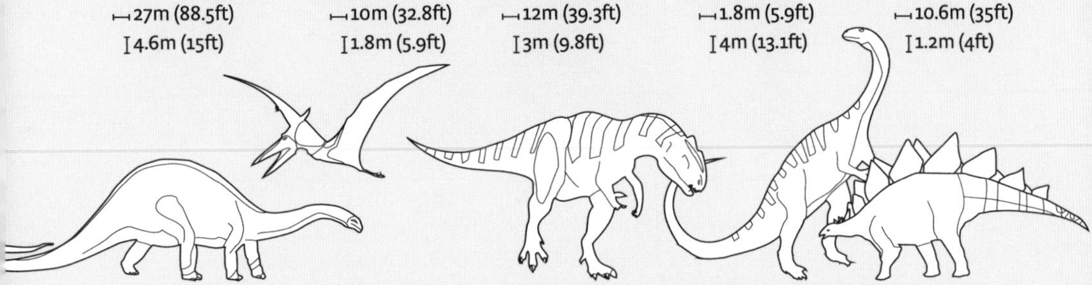

[7] Colorado, Oklahoma, Utah, Wyoming [8] Kansas [9] Colorado, Montana, New Mexico, Oklahoma, S. Dakota, Utah, Wyoming [10] Arizona, [11] Montana

enchantedlearning.com, about.com, wikipedia.org

SNAILS OR **CHOCOLATE?**

Many nations like to define their neighbours by their peculiar taste in food, be it meat, cheese or molluscs. But everyone likes chocolate. Comparing the consumption per capita of the notional national dish in fifteen different countries with their chocolate consumption shows how different (and yet so alike) we all are.

National dish / Consumption per capita per annum

Chocolate consumption

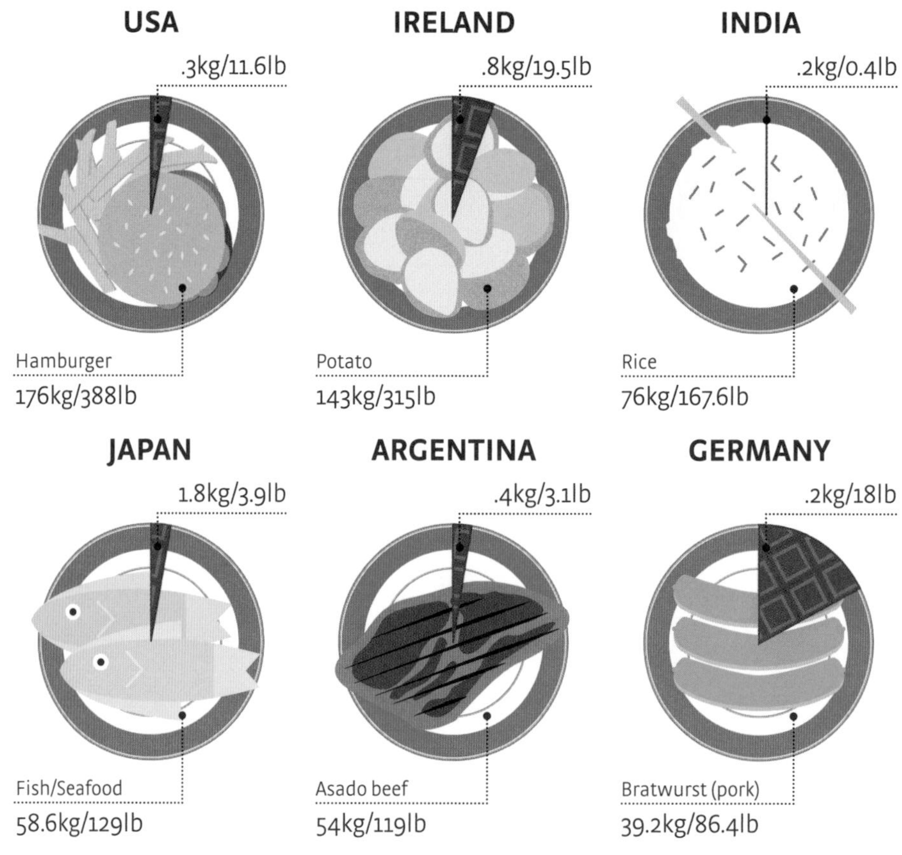

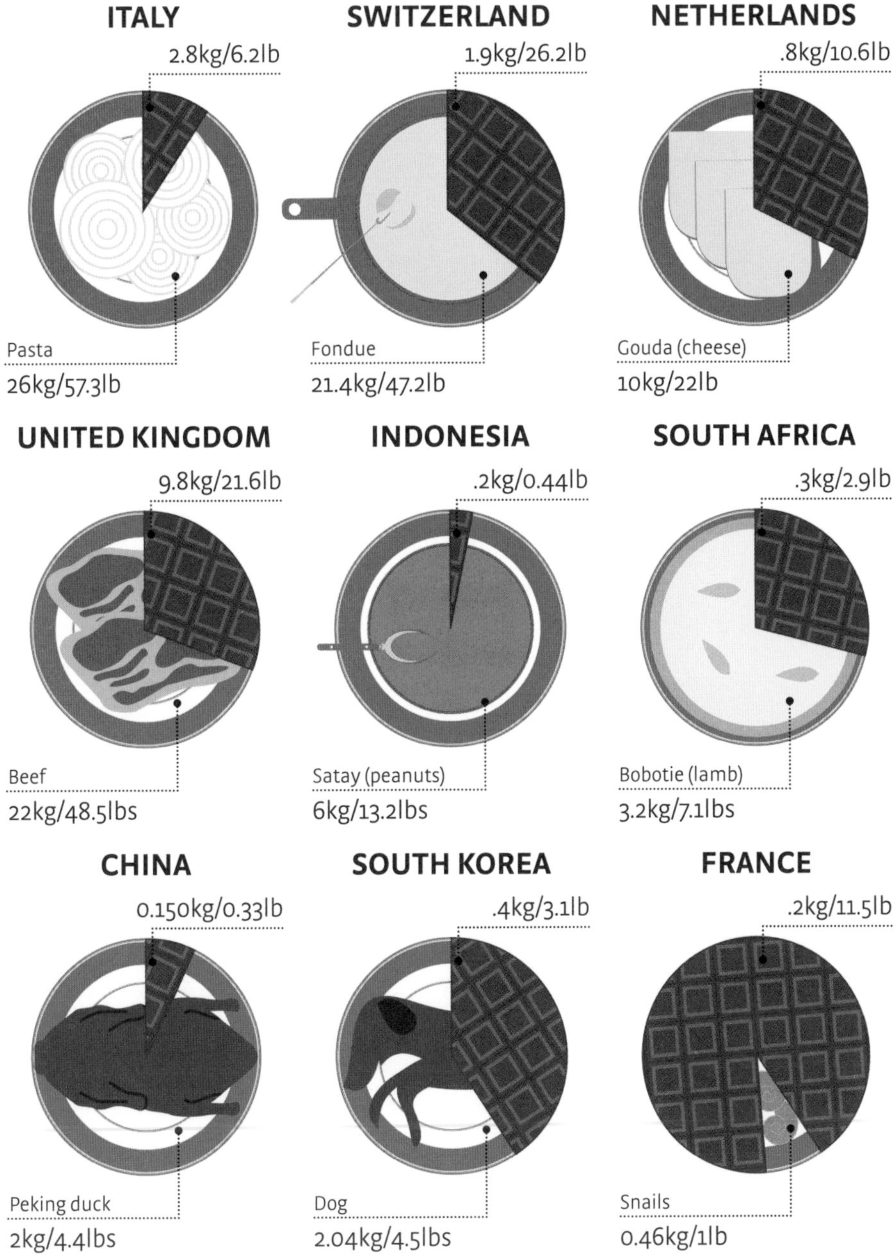

euters.com, bbc.co.uk, internationalpasta.org, economist.com, indexmundi.com, thepigsite.com, pigprogress.net, imv.vu.nl, web-japan.org, the poultrysite.com, meattradenewsdaily.co.uk

EVERYONE WHO EVER LIVED

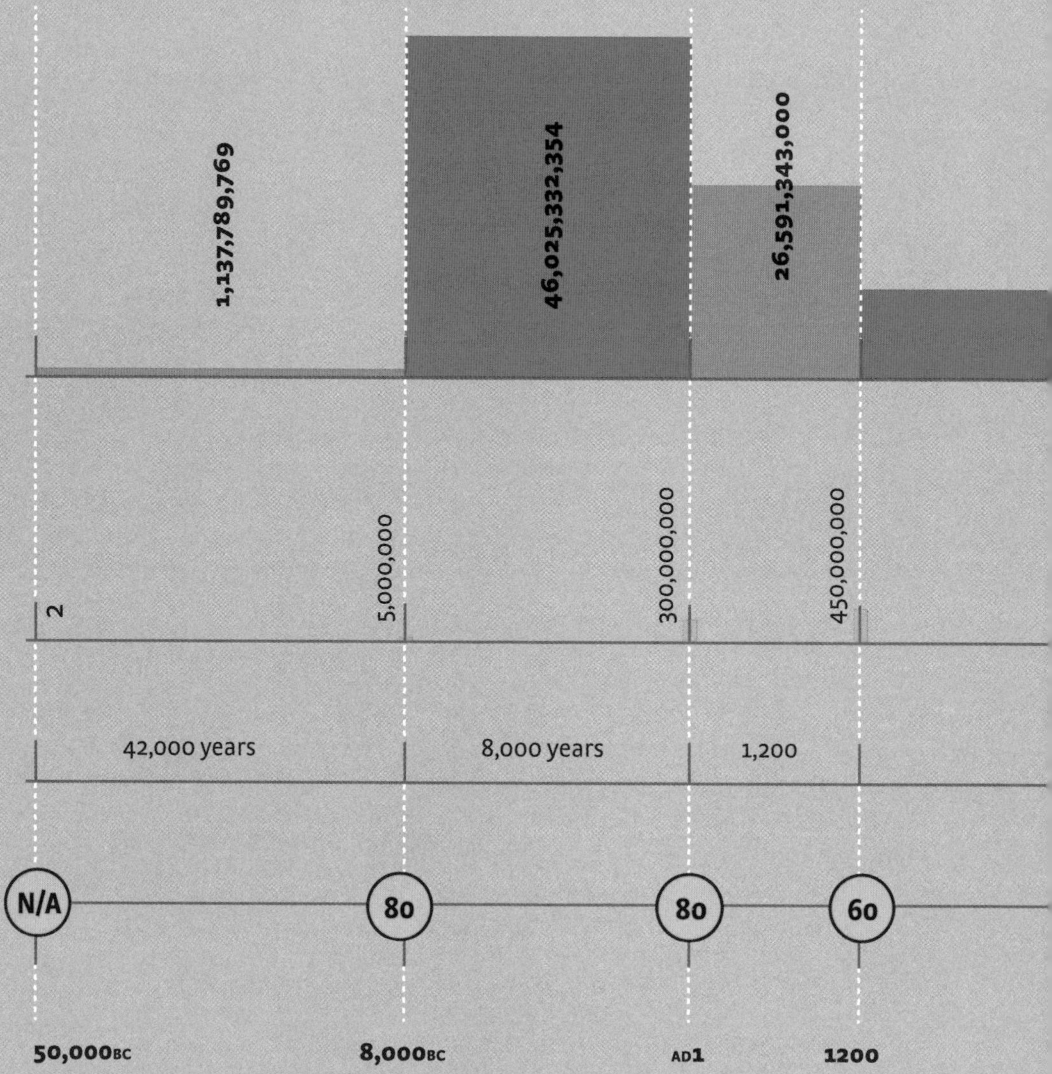

Calculating how many people have lived since the first two humans (let's not call them Adam and Eve) is a difficult business, but the Population Reference Bureau has had a good go at it. Taking into account life expectancy in different eras, plagues, wars, poor diet and lack of healthcare it reckons that there have been more than 107 billion people on earth since 50,000BC.

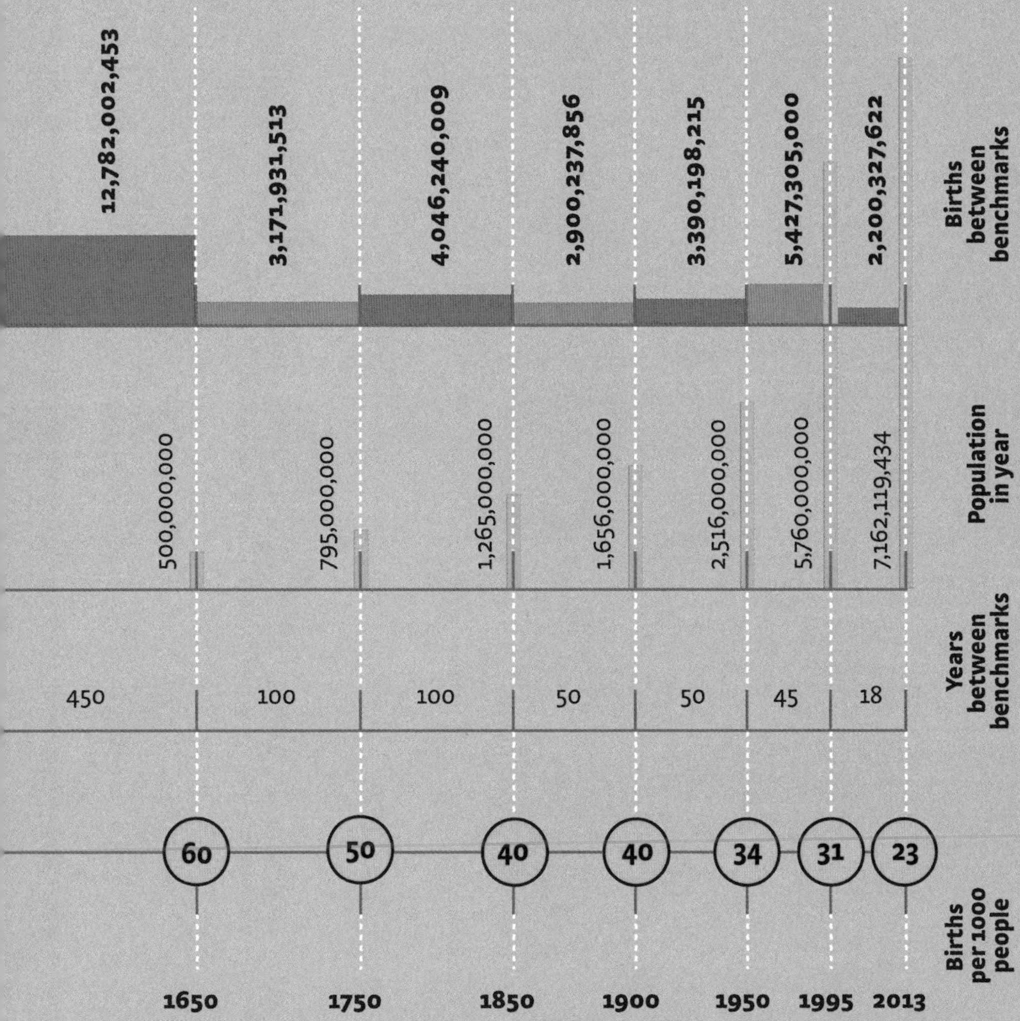

VOYAGER IN MY **POCKET**

When the Voyager space probes launched in 1977 to explore the giant planets and outer solar system they boasted some of the most cutting-edge computer systems ever made. Now, many people carry more computing power than Voyager has in their pockets. The Apple iPhone 5 is over 500 times faster than the Voyager and handles 14 billion instructions per second. Here are some other pieces of everyday technology and their power levels.

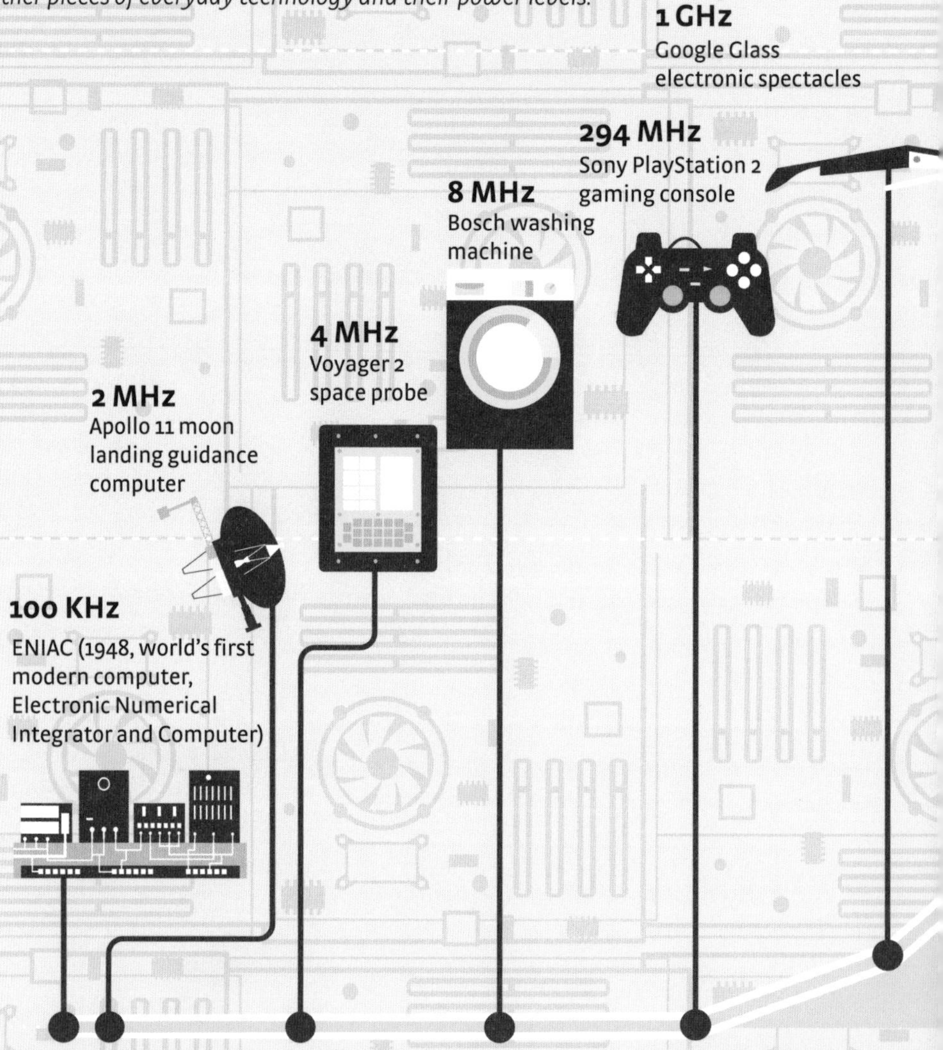

CONVERSION TABLE

1 cycle per second (CPS) = 1 Hertz (Hz)

1000 Hertz (Hz) = 1 kilohertz (KHz)

1000 KHz = 1 megahertz (MHz) is 1000 KHz.

1000 MHz – 1 GHz

1 petaflop = 1 billion floating-point operations per second (there is no conversion between petaflops and Hz).

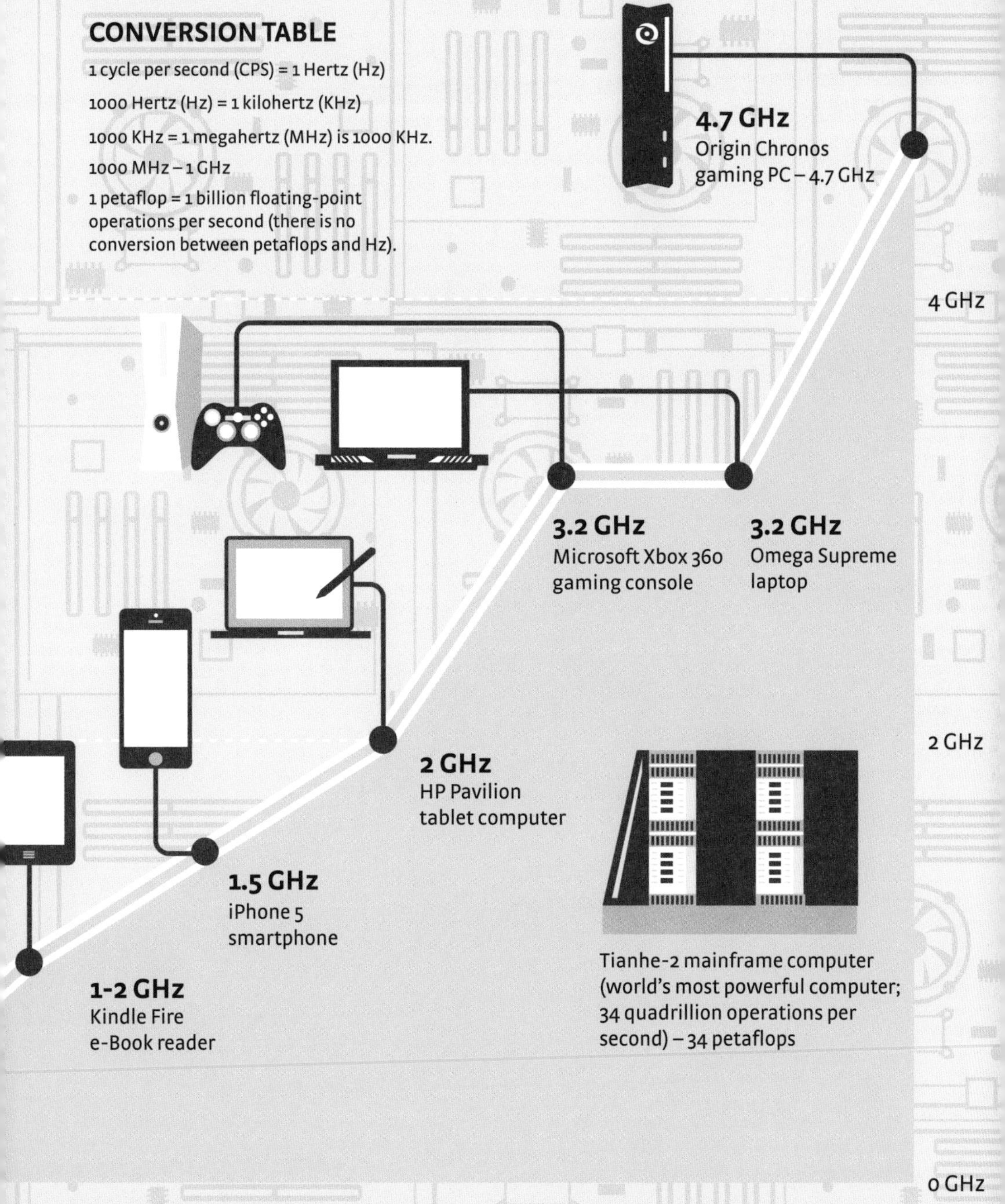

AT THE **EDGE** OF THE **WORLD**

When the first space tourists begin their ascent, these are the stages of the Earth's atmosphere that they'll go through.

DISTANCE FROM EARTH

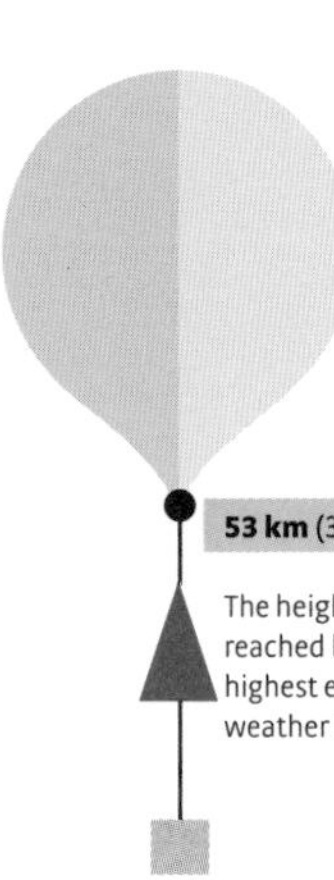

53 km (33 miles)

The height reached by the highest ever weather balloon.

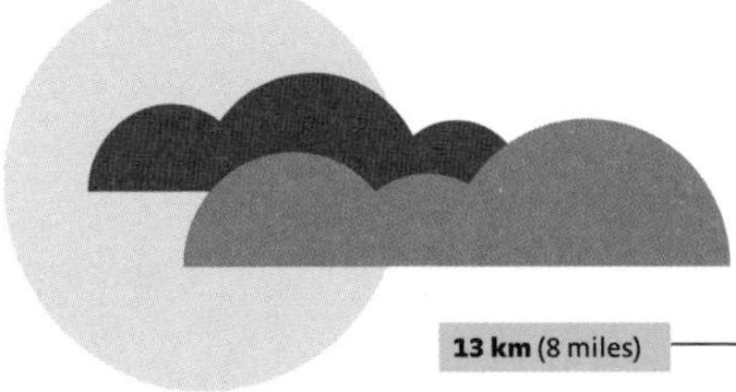

18 km (11 miles)

Without a pressure suit, blood boils at this height.

13 km (8 miles)

Highest clouds (and weather) are far below; all is brilliant sunshine.

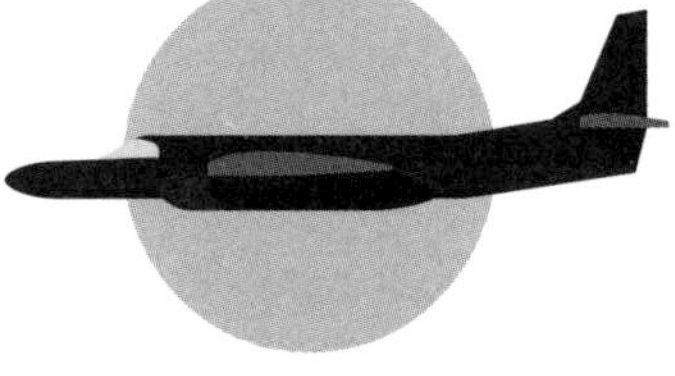

21 km (13 miles)

This is the maximum altitude of a U2 spy plane; 95% of atmosphere, by volume, is below; sky above is black. Curvature of Earth is visible and atmosphere shows on horizon as thin layer of white, light blue and dark blue.

11 km (7 miles)

Commercial airliners fly at this height (well above Mount Everest, at 8.9km/5.5 miles).

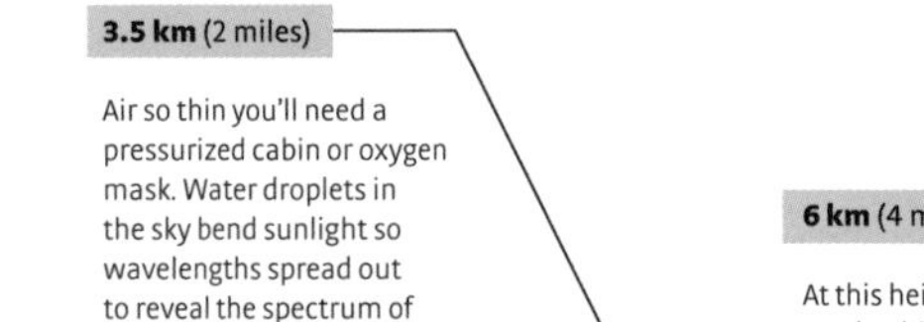

3.5 km (2 miles)

Air so thin you'll need a pressurized cabin or oxygen mask. Water droplets in the sky bend sunlight so wavelengths spread out to reveal the spectrum of colours – or a rainbow. It is an optical effect, so there's no beginning and no end.

6 km (4 miles)

At this height even those used to high altitudes struggle to breathe. The blue sky is determined by molecules of gases and other particles in the air scattering sunlight; wavelengths of blue light are shorter (smaller) than other colours and get scattered more. no beginning and no end.

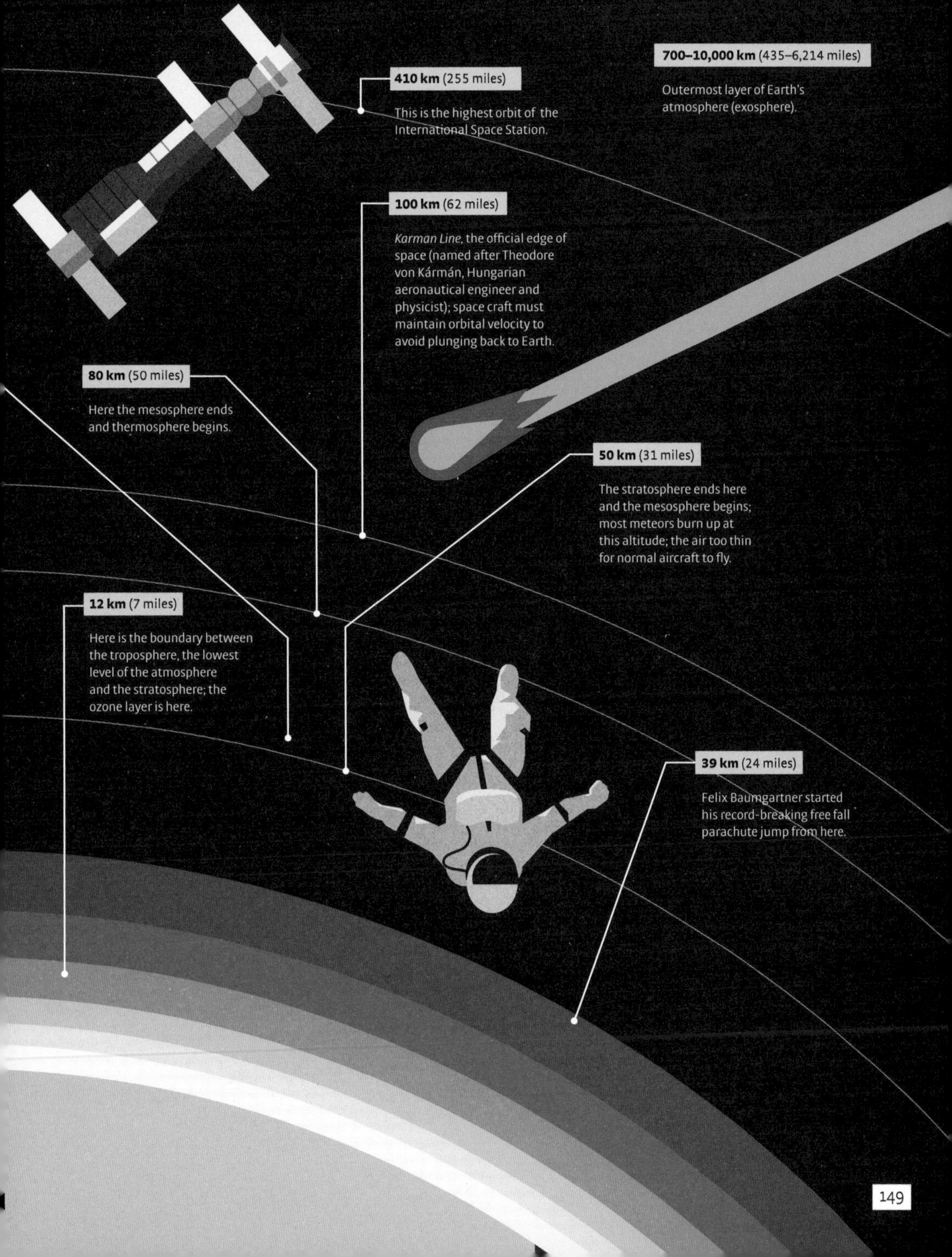
700–10,000 km (435–6,214 miles)
Outermost layer of Earth's atmosphere (exosphere).
410 km (255 miles)
This is the highest orbit of the International Space Station.
100 km (62 miles)
Karman Line, the official edge of space (named after Theodore von Kármán, Hungarian aeronautical engineer and physicist); space craft must maintain orbital velocity to avoid plunging back to Earth.
80 km (50 miles)
Here the mesosphere ends and thermosphere begins.
50 km (31 miles)
The stratosphere ends here and the mesosphere begins; most meteors burn up at this altitude; the air too thin for normal aircraft to fly.
12 km (7 miles)
Here is the boundary between the troposphere, the lowest level of the atmosphere and the stratosphere; the ozone layer is here.
39 km (24 miles)
Felix Baumgartner started his record-breaking free fall parachute jump from here.

THE AIR THAT I **BREATHE**

Air pollution is measured by the number of airborne particles less than 10 micrometers (μm) in diameter per cubic meter of air, they're called PM10s. A PM10 reading of over 35 causes breathing problems in sensitive groups. Above 42 it causes breathing problems in all groups. Readings of 47 or more cause general ill health, and above 65 levels are life-threatening. Apparently New York has far cleaner air than either London or Milan.

☁ =PM10s

POLLUTION BY CITY

City	PM10s
New Delhi, India	198
Islamabad, Pakistan	189
Dhaka, Bangladesh	134
Jeddah, Saudi Arabia	129
Beijing, China	121
Johannesburg, SA	66
Bangkok, Thailand	54
Mexico City, Mexico	52
Milan, Italy	44
Jakarta, Indonesia	43
Paris, France	38
Buenos Aires, Argentina	38
Caracas, Venezuela	37
Moscow, Russia	33
Warsaw, Poland	32
Athens, Greece	31
London, UK	29
Madrid, Spain	26
Berlin, Germany	26
New York, USA	21
Ottawa, Canada	16
Wellington, New Zealand	11
Canberra, Australia	10

MOST AIR POLLUTED PLACES

Ahwaz, Iran	372
Ulan Bator, Mongolia	279
Sanandaj, Iran	254
Ludhiana, India	251
Quetta, Pakistan	251
Kermanshah, Iran	229
Peshawar, Pakistan	219
Gaborone, Botswana	216
Yasuj, Iran	215
Kanpur, India	209

LEAST AIR POLLUTED PLACES

Wollongong, Australia	0.1
Santa Fe, New Mexico, USA	6
Whitehorse, Canada	6
Darwin, Australia	6.9
Kahului-Wailuku, Hawaii, USA	7
Powell River, Canada	8
Springwood, Australia	9
Christchurch, New Zealand	9.31

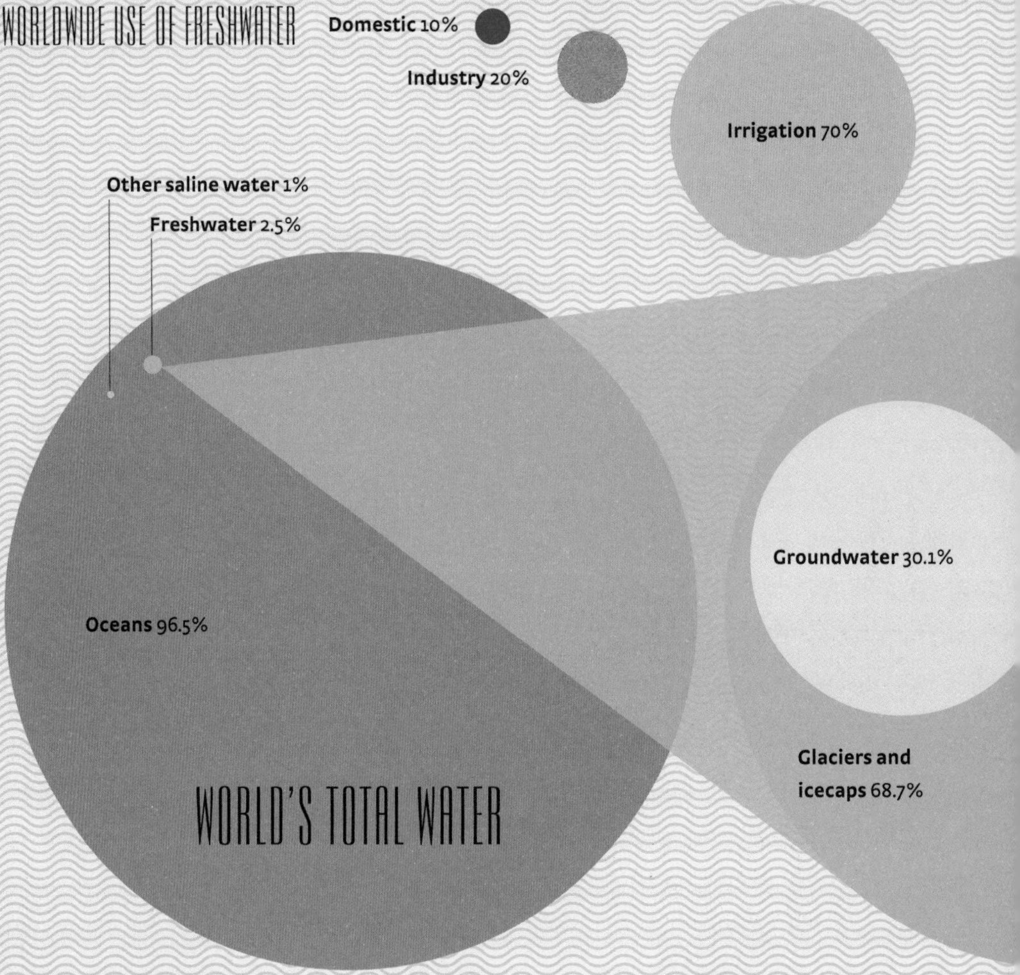

WORLDWIDE USE OF FRESHWATER
Domestic 10%
Industry 20%
Irrigation 70%
Other saline water 1%
Freshwater 2.5%
Groundwater 30.1%
Oceans 96.5%
Glaciers and icecaps 68.7%
WORLD'S TOTAL WATER
WATER NEEDED TO PRODUCE FOOD
(litres per 1kg of food)
Tomato 215
Potato 290
Apple 800
Bread 1,600
Rice 2,500
Chicken 4,300
Butter 5,500
Pork 6,000
Beef 15,500
Chocolate 17,000

WATER, WATER, **EVERYWHERE**

Everyone knows that the Earth's surface is more than 70% covered with water, but where on Earth is all the water, and how much is needed to make a kilo of chocolate?

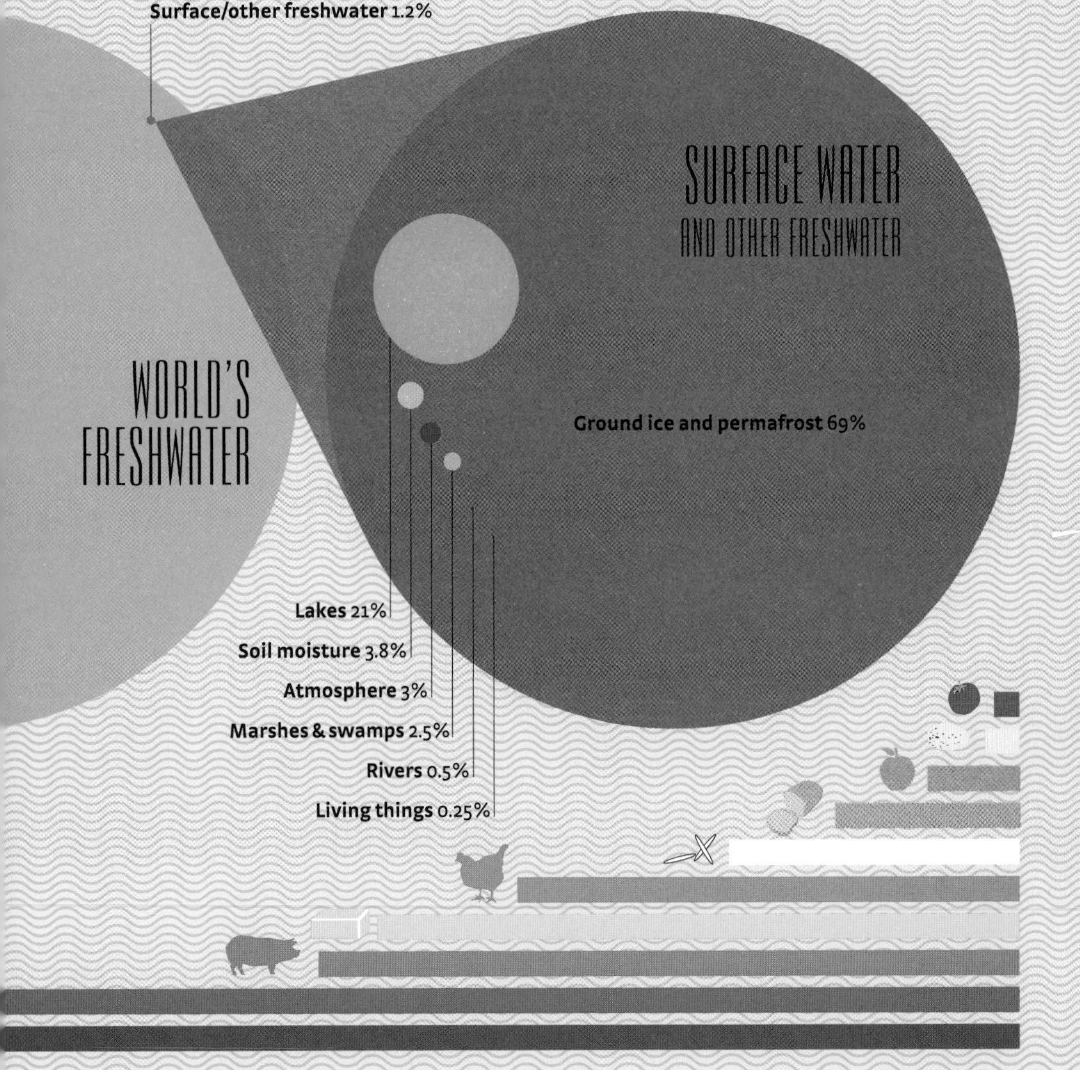

ga.water.usgs.gov, unwater.org, theguardian.com

GANGSTER UNIVERSE

Activities

- Arms Trading
- Blackmail
- Counterfeiting
- Cyber crime
- Child Pornography
- Contract Killing
- Gambling
- Kidnap
- Money Laundering
- Prostitution
- Political Corruption
- People Trafficking
- Drugs
- Extortion
- Fraud
- Robbery

The top gangs in the gangster universe compared via annual turnover, illegal activities and the world-wide reach of their network.

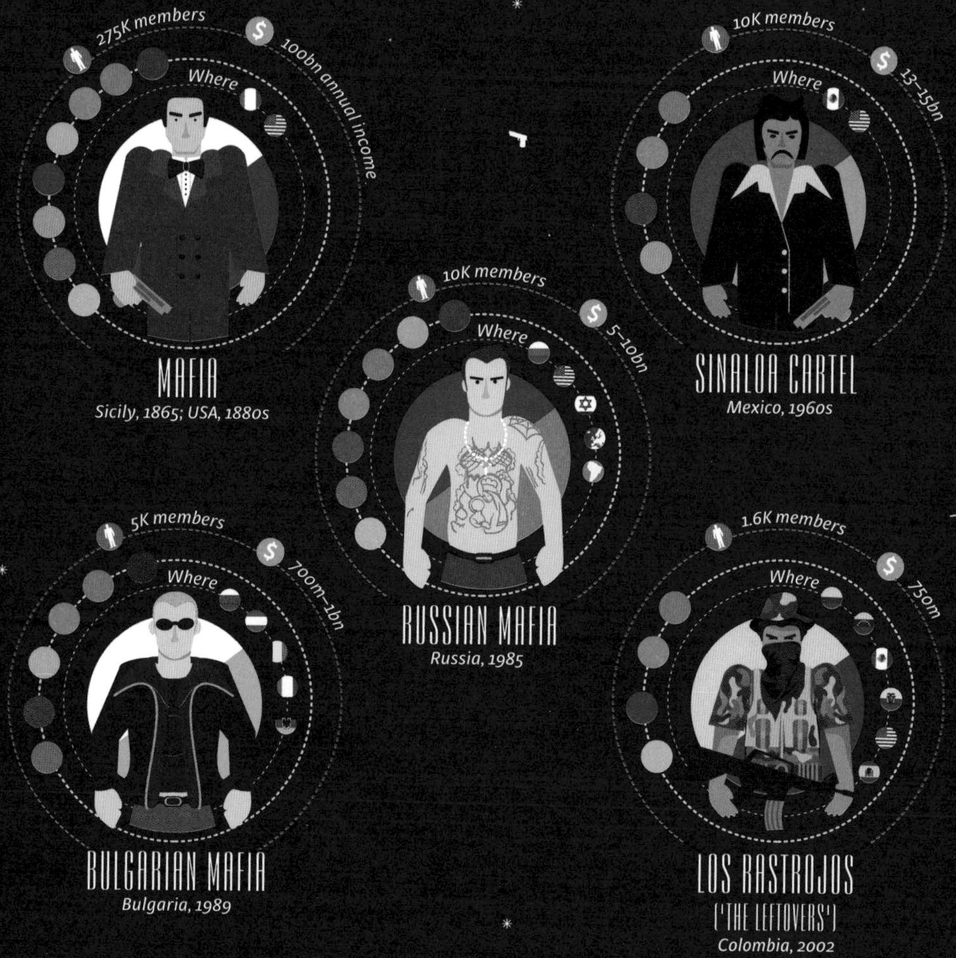

Where

- Japan
- USA, Hawaii
- China, Hong Kong, Taiwan
- Singapore
- The Netherlands
- UK
- Italy, Sicily
- Mexico
- Jamaica
- Colombia
- Russia
- Israel
- Bulgaria
- Belgium
- Germany
- Colombia
- Venezuela
- Ecuador
- Europe
- South America

cia.gov, chinawhisper.com, insightcrime.org, pbs.org, wikipedia.org

WORK, REST **OR PLAY?**

Which are the hardest-working, most sleep-deprived or relaxed and house-proud nations? These Organization for Economic Cooperation and Development nations figures for 18 countries shows how each prefer to mix their weekly work, rest and play.

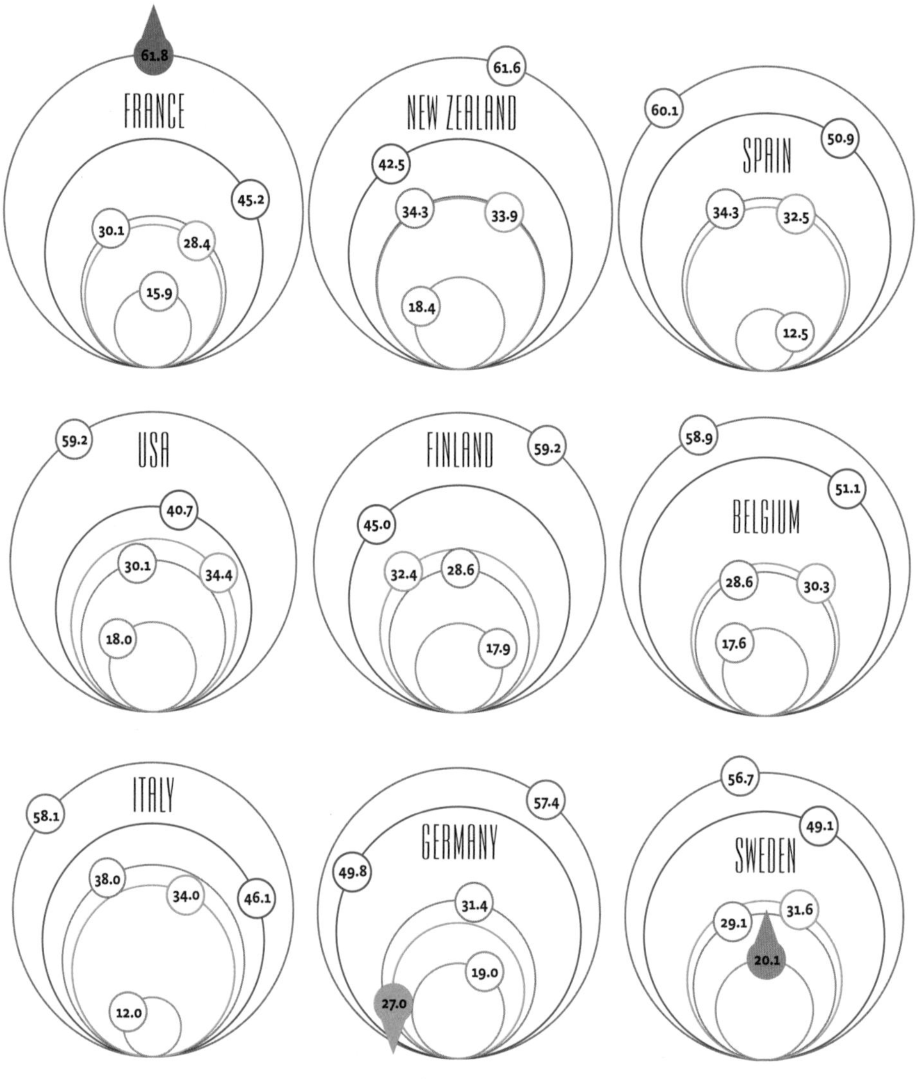

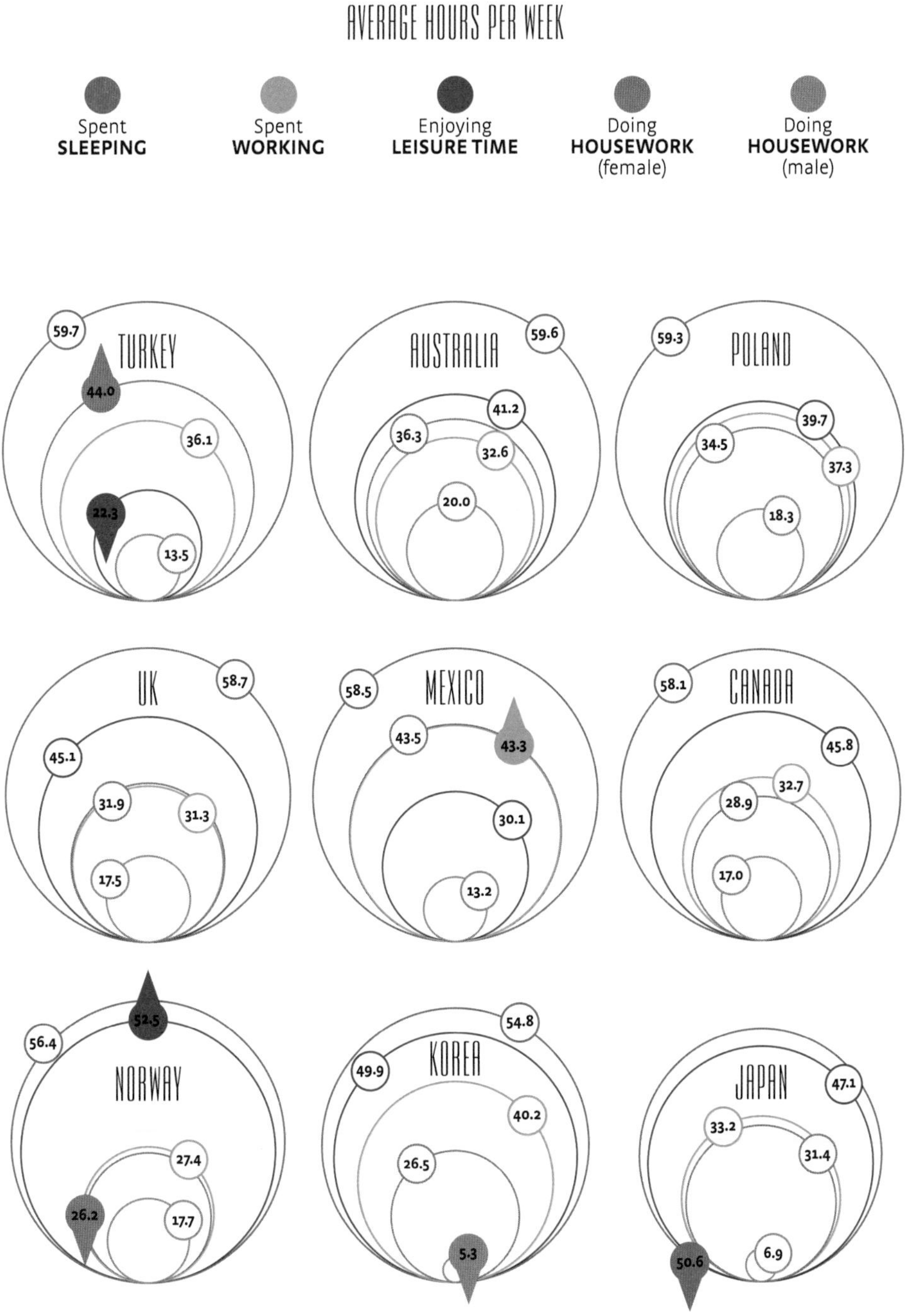
AVERAGE HOURS PER WEEK
Spent SLEEPING
Spent WORKING
Enjoying LEISURE TIME
Doing HOUSEWORK (female)
Doing HOUSEWORK (male)
TURKEY
59.7
44.0
36.1
22.3
13.5
AUSTRALIA
59.6
41.2
36.3
32.6
20.0
POLAND
59.3
39.7
34.5
37.3
18.3
UK
58.7
45.1
31.9
31.3
17.5
MEXICO
58.5
43.5
43.3
30.1
13.2
CANADA
58.1
45.8
32.7
28.9
17.0
NORWAY
52.5
56.4
27.4
26.2
17.7
KOREA
54.8
49.9
40.2
26.5
5.3
JAPAN
47.1
33.2
31.4
6.9
50.6

SIX DEGREES OF SEPARATION: **PETER HIGGS**

The multi-award-winning British theoretical physicist Peter Higgs (b. 1929) has been credited with pioneering work that identified the Higgs Bosun Particle – its existence confirmed by the Large Hadron Collider. It's perhaps unsurprising that he's only 6 steps away from Albert Einstein, but he's also similarly separated from Karl Marx, Earl Grey and 1950s Hollywood heartthrob, James Dean...

1.

EARL GREY
after whom the tea is named, was British Whig PM from 1830–1834, when

HENRY FOX TALBOT
was a Whig MP; he developed a calotype process that aided the work of

LOUIS DAGUERRE
whose photographic images of Paris astounded

SAMUEL MORSE
whose code was transmitted over the first telegraph machine to print on paper, invented by

DAVID E. HUGHES
whose name is given to a Physics prize medal, awarded in 1981 to

TOM W.B. KIBBLE
the British physicist and quantum field theorist, who shared the award with

2.

KARL MARX'S
political theory was a major influence on

CHE GUEVARA
the Argentinian guerrilla activist whose resistance toward

FULGENCIO BATISTA
the President of Cuba in 1959, led to an allegiance with

FIDEL CASTRO
the leader of Cuban rebels who in 1960 as Cuban leader addressed the UN where he met

GOLDA MEIR
then the foreign minister of Israel, whose ambassador from Cuba was

RICARDO WOLF
whose Wolf Foundation prize for Physics was awarded in 2004 to

PETER HIGGS

3.

ALBERT EINSTEIN
signed a letter in 1939 warning about the threat to mankind from H-bombs to

FRANKLIN D. ROOSEVELT
the US President who sanctioned the setting up of the Manhattan Project, run by

LT. GENERAL LESLIE GROVES
who also oversaw the building of the Pentagon, and who appointed as director of the Manhattan Project

J. ROBERT OPPENHEIMER
who recruited physicists from around the world, including

NIELS BOHR
from Denmark, who had been mentor to

OSKAR KLEIN
a Swedish physicist, whose name is given to an annual Memorial Lecture Medal, awarded in 2009 to

4.

JAMES DEAN
the Hollywood film star could recite from memory The Little Prince, written by

ANTOINE DE SAINT-EXUPÉRY
while he was staying at a house built in 1867 by

CORNELIUS H. DELAMATER
a US industrialist who counted as his best friend

JOHN ERICSSON
A Swedish inventor who designed armed battleships and who had collaborated with

ALFRED NOBEL
the inventor of dynamite, whose annual Prize for Physics was shared in 2013 by

FRANCOIS ENGLERT
the Belgian physicist, and

Credits

Produced by Essential Works Ltd
essentialworks.co.uk

Essential Works

Art Director: Gemma Wilson
Commissioning Editor: Mal Peachey
Editor: Fiona Screen
Researchers: Naomi Barton, George Edgeller, Phil Hunt, Jane Mosely, Maria Ines Pinheiro, Kimberley Simpson, Renske Start, Jackie Strachan, Giulia Vallone, Barney White.
Layout: Louise Leffler

Octopus Books

Editorial Director: Trevor Davies

Assistant Production Manager: Lucy Carter

Designers/Illustrators

Marc Morera Agustí (50–51)
Leticia Ansa (54–55)
Kuo Kang Chen (30–31)
Ian Cowles (36–37)
Barbara Doherty (21, 64–65, 77, 80–81, 94–95, 150–151)
Wayne Dorrington (38)
Jennifer Dossetti (61, 124–125)
Maya Eilam (10–11)
Cristian Enache (114–115, 128–129)
Jacopo Ferretti (40–41)
Dan Geoghegan (46–47, 118–119, 120–121, 122–123, 134–135)
Marco Giannini (39, 48–49, 82–83, 88–89, 100–101, 104–105)
Nick Graves (66–67)
Lorena Guerra (18–19, 26–27, 56–57, 58–59, 96–97, 138–139)
Natasha Hellegouarch (78–79, 154–155, 158–159)
Tomasz Kłosinski (42–43, 44–45)
Stephen Lillie (68–69, 76, 86–87, 106–107)
Mish Maudsley (70–71)
milkwhale.com (108–109, 112–113, 146–147)
Bogdan Rauta (60)
Gaia Russo (62–63, 90–91, 92–93)
Aleksandar Savic (12–13, 34–35, 130–131, 132, 133, 148–149)
Yael Shinkar (14–15, 16–17, 22–23, 72–73, 74–75, 84–85, 140–141, 142–143)
Arnold Skawinski (102–103, 116–117, 144–145)
Kristian Tumangan (52–53, 98–99)
Matt Veal (32–33)
Ryan Welch (20)
Gemma Wilson (24–25, 28–29, 126–127, 152–153, 156–157)
Anil Yanik (110–111, 136–137)

Answers to page 104
Uranus: 25,362km/15,759 miles
Venus: 6,052km/3,760 miles
Titan (Saturn): 2,576km/1,600 miles
Moon (Earth): 1,737km/1,079 miles
Earth: 6,371km/3,959 miles
Mercury: 2,440km/1,516 miles
Ganymede (Jupiter): 2,631km/1,635 miles
Jupiter: 269,911km/43,440 miles
Callisto (Jupiter): 2,410km/1,497 miles
Mars: 3,390km/2,106 miles
Europa (Jupiter): 1,561km/970 miles
Pluto: 1,184km/736 miles